TOPICS IN GROUP THEORY AND COMPUTATION

TOPICS IN GROUP THEORY AND COMPUTATION

Proceedings of a Summer School held at University College, Galway, under the auspices of the Royal Irish Academy from 16th to 21st August, 1973

Edited by

MICHAEL P. J. CURRAN
Department of Mathematics
University College
Galway, Ireland

Summer School on Group theory and Computation,
" Galway, Ireland, 1973

1977

ACADEMIC PRESS
London New York San Francisco
A Subsidiary of Harcourt Brace Jovanovich, Publishers

ACADEMIC PRESS INC. (LONDON) LTD.
24/28 Oval Road
London NW1

United States Edition published by
ACADEMIC PRESS INC.
111 Fifth Avenue
New York, New York 10003

Library of Congress Catalog Card Number: 76-016962
ISBN: 0-12-200150-8

Text set in 10/11 pt IBM Press Roman printed by photolithography, and bound in Great Britain at The Pitman Press, Bath

Participants

MR. JURGEN ACKVA, *Gesellschaft fur Mathematik und Datenterarbeitung, Bonn, West Germany.*

DR. JAMES G. ATKINSON, *Government Communications Headquarters, Cheltenham, England.*

MR. JOHN CLIFFORD AULT, *Department of Mathematics, University of Leicester, England.*

DR. KHALIO BENABDALLAH, *Department of Mathematics, Universite de Montreal, Quebec, Canada.*

MR. J. BUHLER, *Department of Mathematics, Harvard University, U.S.A.*

MR. COLIN CAMPBELL, *Mathematical Institute, St. Andrews, Scotland.*

PROFESSOR CYRIL CARTER, *Trent University, Peterborough, Ontario, Canada.*

DR. A. CHRISTOFIDES, *Department of Mathematics, University College, Galway, Ireland.*

MR. G. CLARKE, *Department of Mathematics, University College, Galway, Ireland.*

MR. M. P. J. CURRAN, *Department of Mathematics, University College, Galway, Ireland.*

DR. MARTIN DUNWOODY, *Mathematics Institute, University of Warwick, England.*

MR. G. ENRIGHT, *Department of Mathematics, University College, Galway, Ireland.*

MRS. WALTRAUD FELSCH, *Lehrstuhl D fur Mathematik, Aachen, West Germany.*

DR. LESLIE FLETCHER, *Department of Mathematics, University of Salford, Lancs., England.*

DR. F. GAINES, *Department of Mathematics, University College, Dublin, Ireland.*

MR. ANTHONY GARDINER, *Department of Mathematics, Royal Holloway College, Englefield Green, Surrey, England.*

MR. EBERHARD GELLER, *Gesellschaft fur Mathematic, Bonn, D5205 St. Augustin, 1 Schloss Birlinghoven, West Germany.*

DR. LEONHARD GERHARDS, *Gesellschaft fur Mathematik, D5205 St. Augustin, 1 Schloss Birlinghoven, West Germany.*

DR. EDNA GROSSMAN, *Thomas J. Watson Research Centre, New York, U.S.A.*

MR. RICHARD HARTLEY, *University of Toronto, Ontario, Canada.*

PROFESSOR KURT A. HIRSCH, *Department of Mathematics, Queen Mary College, Mile End Road, London E.1., England.*

DR. CHRISTOPHER H. HOUGHTON, *University College, Cardiff, Wales.*

DR. DAVID JOHNSON, *Department of Mathematics, University of Nottingham, England.*

PROFESSOR KEITH JOSEPH, *Wayne State University, Detroit, Michigan, U.S.A.*

DR. HELMUT JURGENSEN, *Mathematisches Seminar der Christian-Albrechts, Universitat Kiel, Olshausenstrasse, 40–60, West Germany.*

PROFESSOR RAY KEOWN, *Department of Mathematics, University of Arkansas, Fayetteville, Arkansas, U.S.A.*

DR. THOMAS LAFFEY, *Department of Mathematics, University College, Dublin, Ireland.*

MR. P. LANGSTON, *Department of Mathematics, Harvard University, U.S.A.*

DR. CHARLES LEEDHAM-GREEN, *Department of Mathematics, Mile End Road, London, E.1., England.*

DR. JOHN LENNOX, *Department of Mathematics, University College, Cardiff, Wales.*

DR. HANS LIEBECK, *Department of Mathematics, The University of Keele, Staffs., England.*

DR. WOLFGANG LINDENBERG, *Gesellschaft fur Mathematik, Bonn, D5205 St. Augustin 1, Schloss Birlinghoven, Germany.*

PROFESSOR DONALD LIVINGSTONE, *Department of Pure Mathematics, The University of Birmingham, England.*

PROFESSOR J. MCMAHON, *Department of Mathematics, St. Patrick's College, Maynooth, Ireland.*

DR. SPYROS MAGLIVERAS, *State University of New York, College of Arts and Science at Oswego, New York 13126, U.S.A.*

DR. JOHN MCDERMOTT, *Department of Mathematics, The University, Newcastle-upon-Tyne, England.*

DR. T. P. MCDONOUGH, *Department of Mathematics, University College of Wales, Aberystwyth, Wales.*

DR. SUSAN MCKAY, *Department of Mathematics, Queen Mary College, Mile End Road, London E.1., England.*

MR. DAVID MCQUILLAN, *School of Mathematics, Trinity College, Dublin, Ireland.*

PROFESSOR ROBERT MOODY, *Department of Mathematics, University of Saskatchewan, Saskatoon, Canada.*

PROFESSOR W. MOSER, *Department of Mathematics, McGill University, Montreal P.Q., Canada.*

MR. BENJAMIN MWENE, *Department of Pure Mathematics, The University of Birmingham, England.*

DR. M. L. NEWELL, *Department of Mathematics, University College, Galway, Ireland.*

DR. I. O. MUIRCHEARTAIGH, *Department of Mathematics, University College, Galway, Ireland.*

DR. JOSEPH OPPENHEIM, *Department of Mathematics, California State University, San Francisco, U.S.A.*

MR. J. O'REILLY, *Department of Mathematics, University College, Galway.*

MR. M. PENK, *Department of Mathematics, Harvard University, U.S.A.*

PROFESSOR RICHARD RASALA, *Visiting at the Mathematics Institute, University of Warwick, England.*

MR. DAVID RODNEY, *Department of Mathematics, Keele University, Staffs., England.*

MR. CHRISTOPHER ROWLEY, *Department of Mathematics, Open University, Walton, Bletchley, Bucks., England.*

DR. PETER ROWLINGSON, *Department of Computer Science, University of Stirling, Scotland.*

DR. ROBERT SANDLING, *Department of Mathematics, The University, Manchester, England.*

DR. BENEDETTO SCHIMEMI, *Seminario Mathematico, Universita 35100 Padova, Italy.*

MR. PETER EDMUND SMITH, *Department of Pure Mathematics and Mathematical Statistics, Cambridge University, England.*

MR. J. SHEIL, *Department of Mathematics, University College, Galway, Ireland.*

PROFESSOR S. J. TOBIN, *Department of Mathematics, University College, Galway, Ireland.*

DR. M. TORRES, *Department of Mathematics, University of Zaragoza, Spain.*

MR. M. VAZQUEZ, *Department of Mathematics, University of Zaragoza, Spain.*

Authors of Invited Papers

JOHN H. CONWAY, *Department of Pure Mathematics and Mathematical Statistics, 16 Mill Lane, Cambridge CB2 1SB, England.*

JOHN LEECH, *Department of Computing Science, University of Stirling, Stirling, Scotland.*

MARSHALL HALL, JR., *Department of Mathematics, California Institute of Technology, Pasadena, California 91109, U.S.A.*

PETER M. NEUMANN, *The Queens College, Oxford OX1 4AW, England.*

Authors of Invited Papers

[illegible]

[illegible]

[illegible]

[illegible]

Contents

Preface

A Summer School on Group Theory and Computation was held under the auspices of the National Committee for Mathematics of the Royal Irish Academy at University College, Galway, from 16th to 21st July, 1973. It brought together approximately 70 participants from 10 countries.

The Summer School was based on five invited speakers and 15 twenty-minute contributed papers. Videotape recordings were made of four invited speakers, Dr. John H. Conway, Professor John Leech, Professor Marshall Hall, Jr. and Dr. Peter Neumann. These have been placed in the Library of the Language Laboratory at University College, Galway and copies may be ordered. This volume contains, in complete form, the papers given by the invited speakers with the exception of "Groups and their Automorphism Groups" by Professor Kurt A. Hirsch, much of which is to be found in Infinite Abelian Groups, Volume II (Academic Press), by L. Fuchs. Not included are the texts of the contributed papers. Abstracts of these were circulated in mimeographed form to all participants at the start of the Summer School.

The success of the 1973 Summer School was due in the first place, of course, to all the participants. However, the whole venture owes much to the generous financial support of the sponsors. The Royal Irish Academy, the President of University College, Galway, Dr. Martin Newell and Bord Failte held receptions for the participants and their associates. I am most grateful to my colleagues, particularly Professor Sean J. Tobin, for their help and encouragement at all times. Finally, the hard work of Mrs. M. Joyce and the staff of the secretariat at University College, Galway, contributed immeasurably to the success of the Summer School.

1st January, 1976. MICHAEL P. J. CURRAN.

1. Computers in Group Theory†

MARSHALL HALL, JR.

Department of Mathematics, California Institute of Technology, Pasadena, California 91109, USA

1. Introduction

Many of the recent results in group theory have been obtained by the use of computers and in a number of cases it is difficult to see how these results could have been obtained without the use of computers. This is particularly true of the construction of some of the newly discovered sporadic simple groups. As this is being written, there is a possible new group discovered by Michael O'Nan of order $460{,}815{,}505{,}920 = 2^9 . 3^4 . 5 . 7^3 . 11 . 19 . 31$.* It may have $L_3(7)$ as a subgroup and if so will have a permutation representation of 245,520 points. Perhaps C. Sims can find this representation since he has already found the representation of the larger Lyons group on 9,606,125 points.

Coset enumeration is a method well suited to computers which finds a permutation representation of a group given by relations. The second section of this paper discusses this technique as a means of characterizing groups containing a class of elements of a certain kind. The third section discusses the use of coset enumeration as a means of constructing a group G from a subgroup H, presumed to be maximal and of known type, together with some element t not in H, providing enough relations may be found connecting t with the relations of H. A very good example of this is the construction of the Higman-McKay-Janko group.

The fifth section discusses briefly some of the things which have been done with permutation and matrix representations of groups.

The sixth section mentions some work which has been done on the Burnside problem and raises questions as to what might be done in the future.

*(*Added in proof*: C. Sims has constructed this group, assuming the existence of a certain automorphism of order 2.)

† This research was supported in part by ONR Contract N00014-67-A-0094-0010 and in part by NSF grant GP 36230X.

2. Coset enumeration. Groups with given types of relations

Coset enumeration is a method for finding a permutation representation for a group G given by generators and relations. The representation will be on the cosets of an appropriately chosen subgroup H. We will have

$$G = H + Hx_2 + \ldots + Hx_n \tag{2.1}$$

and for each generator b the representation of b will be

$$\pi(b) = \begin{pmatrix} H, & Hx_i & \ldots \\ Hb, & Hx_ib & \ldots \end{pmatrix}. \tag{2.2}$$

If $b_1 b_2 \ldots b_m = 1$ is one of the defining relations of G then for each i

$$Hx_i b_1 \ldots b_m = Hx_i \tag{2.3}$$

Also if $b_1 \ldots b_s$ is an element of the subgroup H then

$$Hb_1 \ldots b_s = H \tag{2.4}$$

The basic algorithm for coset enumeration is due to Todd and Coxeter [12]. The cosets are represented by numbers 1, 2, . . . , with 1 representing the subgroup H. If cosets 1, 2, . . . , t have been defined and if b is a generator we define the action of b on coset j by the rule $jb = k$ if this is a consequence of a relation (2.3) or (2.4) and k is one of the cosets already defined. Otherwise we define $jb = t + 1$ as a new coset. In this process we may find that two different numbers in fact represent the same coset. When this "coincidence" is discovered the larger number representing the coset is replaced by the smaller number and this may lead to further identifications. In the process we first use the relations (2.4) determining the subgroup H and then for each i and each relation use (2.3) keeping a record of the permutations for the generators and their inverses to the extent to which they have been defined. When no new cosets can be defined, then the complete permutations for the generators are determined and all relations are satisfied, and we say that the enumeration has closed.

A simple illustration of this for the group $G = \langle a, b \rangle$, where $a^3 = b^3 = 1$, $abab^{-1}a^{-1}b^{-1} = 1$. We shall take $H = \langle a \rangle$. We write the calculations corresponding to (2.3) in columns.

The completed form of this work is

$$\pi(a) = \begin{pmatrix} 1 & 2 & 3 & 4 & 5 & 6 & 7 & 8 \\ 1 & 4 & 6 & 5 & 2 & 7 & 3 & 8 \end{pmatrix}$$

$$\pi(b) = \begin{pmatrix} 1 & 2 & 3 & 4 & 5 & 6 & 7 & 8 \\ 2 & 3 & 1 & 4 & 8 & 5 & 7 & 6 \end{pmatrix}$$

$$\begin{array}{c|cccccccc} & 1 & 2 & 3 & 4 & 5 & 6 & 7 & 8 \\ a & & & & & & & & \\ & 1 & 4 & 6 & 5 & 2 & 7 & 3 & 8 \\ a & & & & & & & & \\ & 1 & 5 & 7 & 2 & 4 & 3 & 6 & 8 \\ a & & & & & & & & \\ & 1 & 2 & 3 & 4 & 5 & 6 & 7 & 8 \end{array} \tag{2.5}$$

	1	2	3	4	5	6	7	8
b	2	3	1	4	8	5	7	6
b	3	1	2	4	6	8	7	5
b	1	2	3	4	5	6	7	8

	1	2	3	4	5	6	7	8
a	1	4	6	5	2	7	3	8
b	2	4	5	8	3	7	1	6
a	4	5	2	8	6	3	1	7
b^{-1}	4	6	1	5	8	2	3	7
a^{-1}	2	3	1	4	8	5	7	6
b^{-1}	1	2	3	4	5	6	7	8

This shows that $H = \langle a \rangle$ has 8 cosets in G and as H is of order 3, G is of order 24. As H is faithfully represented, so is G. G is $SL_2(3)$, the special linear group of degree 2 over the field $GF(3)$. G has a center Z of order 2 generated by $ab^{-1}ab^{-1}$. G/Z is isomorphic to the alternating group A_4 whose defining relations may be taken as $a^3 = b^3 = 1, ab^{-1}ab^{-1} = 1$.

Conway's group .0 [10], the group of automorphisms of the 24-dimensional Leech lattice, contains a class of elements of order 3 with the property that any two elements of this class either (a) commute, (b) generate $SL_2(3)$ or A_4, or (c) generate $SL_2(5)$ or A_5. Here $SL_2(5)$ is of order 120 and modulo a center of order 2 is the alternating group A_5. The group .0 has a center of order 2 and modulo this center the group is a simple group .1. In .1 the class does not have two elements which generate $SL_2(5)$.

From this connection with the Conway group it seems reasonable to investigate groups with a class of elements of order 3 such that any two satisfy one of the following sets of relations:

$$(1)\ a^3 = b^3 = 1, \quad ba = ab, \quad \langle a, b \rangle \text{ Abelian}$$

$$(2)\ a^3 = b^3 = 1, \quad bab = aba, \quad \langle a, b \rangle = SL_2(3)$$

$$(3)\ a^3 = b^3 = 1, \quad ab^{-1}ab^{-1} = 1 \quad \langle a, b \rangle = A_4$$

$$(4)\ a^3 = b^3 = 1, \quad (ab)^5 = 1 \quad (aba^{-1}b)^2 = 1, \langle a, b \rangle = A_5 \tag{2.6}$$

If we were to include $SL_2(5)$, its defining relations are

$$(5)\ a^3 = b^3 = 1, \quad babab = ababa, \quad b^{-1}ab^{-1}ab^{-1} = ab^{-1}ab^{-1}a \tag{2.7}$$

In order to describe these groups it is necessary to consider those generated by a small number of elements, and this has been done on a computer using coset enumeration. Aschbacher and Hall [1] determined all such groups in which only relations 1, 2, or 3 arose. Here it is convenient to use the notation $a \sim b$ for the relations $a^3 = b^3 = 1$, $bab = aba$.

For three generator groups $G = \langle a, b, c \rangle$, in which $a^3 = b^3 = c^3 = 1$ and $a \sim b$, the following cases arise:

(1) $ca = ac$, $cb = bc$. $\langle a, b, c \rangle = \langle a, b \rangle \times c$.

(2) $ca = ac$, $c \sim b$, $G = \langle a, b, c \rangle$. G is of order 648 and putting $h = a^{-1}c$, G has a normal subgroup $H = \langle h, b^{-1}hb \rangle$ of exponent 3 and order 27 and $G/H = \langle a, b \rangle$. In G there are 12 groups conjugate to $\langle a \rangle$.

(3) $c \sim a$, $c \sim b$, $c \sim a^{-1}ba$, $c \sim b^{-1}ab$. Here G has order $768 = 2^8.3$ and G has 16 subgroups conjugate to $\langle a \rangle$.

(4) $c \sim a$, $c \sim b$, $c \sim a^{-1}ba$, $c^{-1} \sim b^{-1}ab$. These relations make G collapse so that $G = 1$.

(5) $c \sim a$, $c \sim b$, $c^{-1} \sim a^{-1}ba$, $c^{-1} \sim b^{-1}ab$. Here G is of order 6048 and is isomorphic to the simple group $U_3(3)$. G contains 28 subgroups conjugate to $\langle a \rangle$.

If $\langle a, b, c \rangle = U_3(3)$ careful analysis shows that a fourth generator d in the class D of conjugates of $\langle a \rangle$ must necessarily commute with at least one of the 28 conjugates of $\langle a \rangle$ in $U_3(3)$ if $\langle a, b, c, d \rangle = G$ does not collapse to the identity. If d commutes with exactly one of these 28, which we may take to be $\langle a \rangle$, then for the 27 remaining $\langle x_i \rangle$'s conjugate to $\langle a \rangle$ we may suppose $d \sim x_i$ whenever $a \sim x_i$. In this case $G = \langle a, b, c, d \rangle$ has a normal subgroup $K = \langle (a^{-1}d)^g \rangle$, $g \in G$, K is a 3-group, and $G/K \cong \langle a, b, c \rangle = U_3(3)$.

If d commutes with more than one but not all 28 conjugates, then we may take $da = ad$, $db = bd$ and $d \sim c$. In this case G has a center of order 36 and modulo the center is the simple group $U_4(3)$.

With these preliminaries Aschbacher and Hall proved the following theorem.

Theorem. *Let* G *be a finite group generated by a conjugacy class* D *of subgroups of order* 3, *such that for any pair of noncommuting subgroups* A *and* B *in* D, *the group generated by* A *and* B *is isomorphic to* $SL_2(3)$ *or* A_4. *Assume* G *contains no nontrivial solvable normal subgroup. Then* G *is isomorphic to a symplectic group* $Sp_{2n}(3)$, *a unitary group* $U_n(3)$ *or a projective generalized unitary group* $PGU_n(2)$.

For cases involving an A_5, using the results of Aschbacher and Hall, and of Assion [2], Stellmacher [24] has completed the classification. His final result is given in the following theorem:

Theorem (Assion and Stellmacher).

Let G *be a finite group with the following properties*:

(1) G *is generated by a conjugate class* D *of elements of order* 3. *Two noncommuting elements of* D *generate a subgroup isomorphic to* A_4, A_5 *or* $SL_2(3)$.

(2) *There exist elements in* D *which generate a subgroup isomorphic to* A_5.
(3) $0_2(G) = Z(G) = 1$.

Then G *is isomorphic to* $Sp_{2n}(2)$ *for* $n \geqslant 3$, $0^+_{2n}(2)$ *or* $0^-_{2n}(2)$ *for* $n \geqslant 3$, A_n *for* $n \geqslant 5$, *the Hall-Janko group,* $G_2(4)$, *the Suzuki group or the Conway group* .1. *The conjugacy class* D *is uniquely determined except when* G *is isomorphic to* A_6 *or* $0^+_8(2)$.

These results depend very heavily on certain initial cases established by coset enumeration. Some of these were carried out by Assion and Stellmacher, and some by John McKay. Two cases of these enumerations are worth noting. It is suggestive to represent the relations by diagrams in which we have X•—•Y to represent $X^3 = Y^3 = 1, XYX = YXY$ (i.e. $X \sim Y$) where $\langle X, Y\rangle = SL_2(3)$, or $\langle X, Y\rangle = A_4$ if further relations collapse $XY^{-1}XY^{-1}$ to the identity. We use X═Y to represent $X^3 = Y^3 = 1, (XYX^{-1}Y)^2 = 1, (XY)^5 = 1$ where $\langle X, Y\rangle = A_5$. If $X \sim Y$ the four elements conjugate to X in $\langle X, Y\rangle$ are $X, X^{-1}YX, Y^{-1}XY, Y$.

$$G = \langle A, B, C\rangle = HJ$$

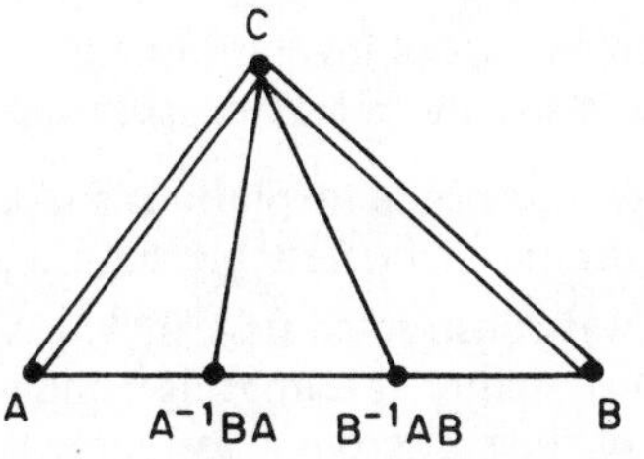

(2.8)

$$H = \langle A, C, B^{-1}CB\rangle, \quad [G:H] = 315$$

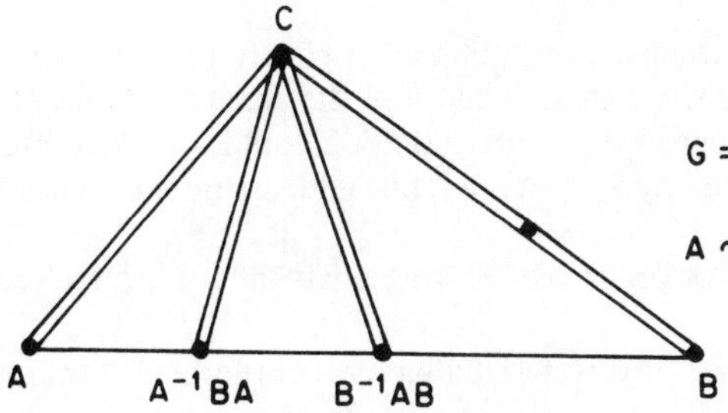

$G = \langle A,B,C\rangle = G_2(4)$

$A \sim C^{-1}B^{-1}CBC$ (2.9)

$$H = \langle C^{-1}AC, B^{-1}CB, CAC^{-1}, CBC^{-1}\rangle, [G:H] = 416$$

In these coset enumerations it is important to choose the subgroup H judiciously. In these cases as the group is a simple group, H is chosen to be a maximal proper subgroup. No clear cut rule is known for choosing such a subgroup. The relations as given here are due to John McKay.

3. Coset enumeration. Construction of simple groups

In 1967 Zvonimir Janko announced the possible existence of two new finite simple groups in which an involution z (an element of order 2) in the center of a Sylow 2-subgroup had as its centralizer a particular group K of order 1920. Here K contains a normal subgroup E of order 2^5 which is the central product of a quaternion group with a dihedral group of order 8. Precisely E is a factor group of index 2 in the direct product $Q \times D_8$ of a quaternion group Q and a dihedral group D_8. Both Q and D_8 have a center of order 2 and a factor group which is the elementary group of order 4. If we add to the relations for Q and D_8 in their direct product the further relation that their centers are equal this is the central product E. Here K is the splitting extension of E by a factor group isomorphic to A_5. Janko was able to show that a group G, if it existed, was necessarily of one of two types.

First type: G has a single class of involutions and is of order 50,232,960 = $2^7 . 3^5 . 5 . 17 . 19$ and has a specific character table and other specific properties.

Second type: G has two classes of involutions and is of order 604,800 = $2^7 . 3^3 . 5^2 . 7$ and has a specific character table and other specific properties.

The second of these was constructed first, in August 1967, by the author making use of the Titan computer at Cambridge, and more will be said of this later. The construction of the first group was a more troublesome task and was carried out by Graham Higman and John McKay [19] in 1969 by a brilliant use of coset enumeration carried out independently by John McKay on the Atlas computer at Chilton and by M. J. T. Guy on the Titan computer at Cambridge.

There were errors in both character tables as originally proposed by Janko. Indeed the author received word that, because of the character table, the group (now called the Hall-Janko group) of order 604,800 could not exist after it had been constructed. The errors were corrected, but added a note of confusion to the early work.

If Janko's simple group of order 50,232,960 exists then it has the (corrected) character Table 3.1.

In this table a class of elements is designated by its order and if there is more than one class of elements with the same order subscripts are used to distinguish the different classes. Thus there are two classes of elements of order 3 designated by 3_1 and 3_2. Above an element x is written the order of its centralizer so that

TABLE 3.1 G order $50{,}232{,}960 = 2^7.3^5.5.17.19$

$C(x)$	$\|G\|$	1920	96	8	24	12	30	30	10	10	15	15	17	17	19	19	1080	243	27	27	27
Element x	1	2	4	8	6	12	5_1	5_2	10_1	10_2	15_1	15_2	17_1	17_2	19_1	19_2	3_1	3_2	9_1	9_2	9_3
χ_1	1	1	1	1	1	1	1	1	1	1	1	1	1	1	1	1	1	1	1	1	1
χ_2	323	3	3	−1	0	0	B	A	B	A	B	A	0	0	0	0	8	−1	−1	−1	−1
χ_3	323	3	3	−1	0	0	A	B	A	B	A	B	0	0	0	0	8	−1	−1	−1	−1
χ_4	324	4	4	0	1	1	−1	−1	−1	−1	−1	−1	1	1	1	1	9	0	0	0	0
χ_5	1938	2	−2	0	−1	1	A	B	$-A$	$-B$	A	B	0	0	0	0	3	−6	0	0	0
χ_6	1938	2	−2	0	−1	1	B	A	$-B$	$-A$	B	A	0	0	0	0	3	−6	0	0	0
χ_7	646	−10	2	0	−1	−1	$2A$	$2B$	0	0	$-A$	$-B$	0	0	0	0	7	−2	1	1	1
χ_8	646	−10	2	0	−1	−1	$2B$	$2A$	0	0	$-B$	$-A$	0	0	0	0	7	−2	1	1	1
χ_9	2754	−14	−2	0	1	1	−1	−1	1	1	−1	−1	0	0	−1	−1	9	0	0	0	0
χ_{10}	816	−16	0	0	2	0	1	1	−1	−1	1	1	0	0	−1	−1	6	6	0	0	0
χ_{11}	3078	−10	2	0	−1	−1	−2	−2	0	0	1	1	1	1	0	0	−9	0	0	0	0
χ_{12}	2432	0	0	0	0	0	2	2	0	0	−1	−1	1	1	0	0	−16	2	−1	−1	−1
χ_{13}	1140	20	−4	0	−1	−1	0	0	0	0	0	0	1	1	0	0	15	6	0	0	0
χ_{14}	1215	15	3	1	0	0	0	0	0	0	0	0	C	D	−1	−1	0	0	0	0	0
χ_{15}	1215	15	3	1	0	0	0	0	0	0	0	0	D	C	−1	−1	0	0	0	0	0
χ_{16}	1615	15	−1	−1	3	−1	0	0	0	0	0	0	0	0	0	0	−5	−5	1	1	1
χ_{17}	85	5	1	−1	−1	1	0	0	0	0	0	0	0	0	E	F	−5	4	1	1	1
χ_{18}	85	5	1	−1	−1	1	0	0	0	0	0	0	0	0	F	E	−5	4	1	1	1
χ_{19}	1920	0	0	0	0	0	0	0	0	0	0	0	−1	−1	1	1	0	3	R	S	T
χ_{20}	1920	0	0	0	0	0	0	0	0	0	0	0	−1	−1	1	1	0	3	S	T	R
χ_{21}	1920	0	0	0	0	0	0	0	0	0	0	0	−1	−1	1	1	0	3	T	R	S

$A = \frac{1+\sqrt{5}}{2}$ $B = \frac{1-\sqrt{5}}{2}$ $C = \frac{-1+\sqrt{17}}{2}$ $D = \frac{-1-\sqrt{17}}{2}$ $E = \frac{-1+\sqrt{-19}}{2}$ $F = \frac{-1-\sqrt{-19}}{2}$

$R = -\rho^2+\rho^4+\rho^5-\rho^7$ $S = -\rho^8+\rho^7+\rho^2-\rho$ $T = -\rho^5+\rho+\rho^8-\rho^4$ ρ a primitive 9th root of unity for example $c(x) = 1080$ if $x = 3_1$ and $c(x) = 243$ if $x = 3_2$. For any matrix representation ρ of a finite group G over the complex field the character $\chi(x)$ of $x \in G$ is the trace of the matrix $\rho(x)$. Conjugate elements have the same trace. Every representation (over the complex field) has an essentially unique decomposition as the sum of irreducible representations $\rho_1, \ldots, \rho_m$ and Table 3.1 gives the values of the corresponding characters χ for representatives of the conjugate classes.

A transitive permutation representation of a group G on n points may be considered as a representation of G by n by n matrices. For $x \in G$, if

$$\pi(x) = \begin{pmatrix} 1 & \ldots r & \ldots n \\ b_1 & \ldots b_r & \ldots b_n \end{pmatrix}$$

then in the matrix $\rho(x) = [a_{ij}]$ $i, j = 1, \ldots, n$ but $a_{r,b_r} = 1, r = 1, \ldots, n$ and $a_{ij} = 0$ otherwise. The character $\varphi(x)$ of $\rho(x)$ is the number of points fixed by the permutation $\pi(x)$.

A permutation character φ satisfies the following properties:

(*a*) φ contains the principal character (here χ_1) just once;
(*b*) $\varphi(1)$ divides $|G|$;
(*c*) $\varphi(x)$ is a nonnegative rational integer for all x in G;
(*d*) $\varphi(1)$ divides $h_i\varphi(x)$, if x belongs to a class containing h_i elements;
(*e*) $\varphi(x) \leqslant \varphi(x^k)$ for all x in G and all integers k.

If φ is the character of the permutation representation of G on the $n = \varphi(1)$ cosets of a subgroup H then if H contains w conjugates of x, since x is in $\varphi(x)$ conjugates of H, we have $nw = h_i\varphi(x)$ if x has h_i conjugates, by counting occurrences of conjugates of x in conjugates of H. This is condition (*d*). It is not known whether there are other conditions on a permutation character.

Using a computer and the character table (Table 3.1) John McKay found the character φ of smallest degree for a possible permutation character of G. This smallest degree was 6156 and the character φ is

$$\varphi = \chi_1 + \chi_2 + \chi_3 + \chi_4 + \chi_{13} + \chi_{14} + \chi_{15} + \chi_{16}. \tag{3.2}$$

The order of the subgroup H would be $8160 = 2^5 . 3 . 5 . 17$. This order is twice that of the group $L_2(16)$ and this suggests that H is $L_2(16)$ extended by an outer automorphism of order 2. The character table for such a group H is known, and restriction of the characters of Table (3.1) to the assumed H gave no inconsistencies, and so tended to confirm this assumption.

From here on we shall assume that G exists and that it has a subgroup $H = \langle L_2(16), s\rangle$ where $s^2 = 1$ and s induces the outer automorphism of order 2 in $L_2(16)$ by conjugation. A Sylow 2-subgroup V of $L_2(16)$ is elementary Abelian of order 16. From the consideration of the structure of the centralizer of an involution in G it will follow that $L = N_G(V)$ is the split extension of V by $GL(2, 4)$ acting naturally on V, so that $|L| = 2880$. But $N_H(V)$ is only of order 480 so that in G, $H \cap L = N_H(V)$. Hence, as H is a maximal subgroup of G we must have $G = H \cup L$. Now it is not difficult to show that $L = \langle N_H(V), t\rangle$ with an element t such that $t^2 = 1$.

We now seek generators and relations to define G. We find generators and relations for H and then combine the appropriate elements from $N_H(V)$ with t to generate L. These relations alone would define the amalgamated free product of H and L with $N_H(V)$ the amalgamated subgroup. This is, of course, an infinite group and further relations are needed to determine the group G. A further relation can be found by constructing an element not in H which normalizes a Sylow 17-group of H, and as the character table tells us that Sylow 17-normalizer is a Frobenius group of order 136, this forces a specific relation. Of course there might be even further relations required to determine G as a factor group of the free amalgam $H \cup L$, but putting these relations on the Atlas computer at Chilton, John McKay found by coset enumeration that a group G was generated on 6156 cosets of H.

First we have $L_2(16) = \langle a, b, c\rangle$ where

$$a^2 = c^2 = b^{15} = (ac)^3 = (bc)^2 = ab\,ab^{-4}ab^3 = 1 \tag{3.3}$$

Then $H = \langle L_2(16), s\rangle$ where a, b, c, s satisfy the further relations

$$s^2 = (sa)^2 = (sc)^2 = sbsb^{-4} = 1 \tag{3.4}$$

Finally $G = \langle a, b, c, s, t\rangle$ where the following relations involve t:

$$t^2 = (at)^2 = (bt)^3 = b^5tb^{-5}t = (ct)^4s = (b^2st)^3 \tag{3.5}$$

$$(b^{-2}ctb^4ct)^2 = b^2tb^{-1}abtb^{-2}a = 1$$

$$b^{-2}ab^{-3}ctab^2ctb^3ab^3ctactb^7ab^4ct = 1$$

Higman and McKay in fact determine a group G_1 in which G is of index 2, starting from a group $H_1 = \langle L_2(16), u\rangle$ where $u^4 = 1$ induces an outer automorphism of order 4 in $L_2(16)$ and $u^2 = s$. The group G_1 also has 6156 cosets of H_1 and is of rank 7 having suborbit lengths 1, 85, 120, 510, 680, 2040, and 2720. G is of rank 8, the suborbit of length 2720 in G_1 splitting into two orbits of length 1360.

Higman and McKay show that the group G constructed is simple and also that there is a single class of involutions in G and that the centralizer of an involution has the structure given by Janko.

McKay and Wales [22] showed that G has a Schur multiplier of order 3. This means that there is a group G^* with a center $\langle y\rangle$ of order 3 and $G^*/\langle y\rangle$ is isomorphic to G. If we take an element y commuting with a, b, c, s, t and the relations (3.3), (3.4) and all but the last relation of (3.5) but replace the last relation of (3.5) by

$$b^{-2}ab^{-3}ctab^2ctb^3ab^3ctactb^7ab^4ct = y,\ y^3 = 1 \tag{3.6}$$

these define G^*. The group G^* has a matrix representation of degree 18.

4. Calculations of group characters and applications

Different groups may have the same character table. For example, the quaternion group and the dihedral group of order 8 have the same character table and there is considerable duplication in the tables for p-groups. But many groups, for example the alternating and symmetric groups, are uniquely determined by their character tables, and this is true for all known simple groups for which the character tables are also known.

Characters were used quite heavily by the writer in his search for simple groups of order less than one million. The underlying mathematical theory and the computer methods used here are applicable in many other cases. A very important theory due to Richard Brauer [4] is applicable to groups whose order is divisible by exactly the first power of a prime. It rests on deep and difficult theory of modular representation but the results are easy to describe and well suited for numerical analysis.

We suppose that a group G has order g divisible by exactly the first power of a prime p. Then we may write

$$g = (1 + rp)pqv \tag{4.1}$$

where $1 + rp$ is the number of Sylow p-subgroups and pqv is the order of $N_G(P)$. The centralizer of P, $C_G(P) = P \times V$ has order pv and q is the order of the automorphisms induced in P in $N_p = N_G(P)$. Since the automorphisms of a group of order p are a cyclic group of order $p - 1$, we necessarily have $q | p - 1$. Hence in N_p there is an element d such that $d^{-1}cd = c^s$ where $c \to c^s$ is an automorphism of $P = \langle c \rangle$ of order q and $d^q \in V$.

In the Brauer theory there are a number of characters of G, called the principal p-block $B_0(p)$ in the modular theory, consisting of the principal character χ_1 for which $\chi_1(x) = 1$ for every x of G and further characters $\chi_2, \ldots, \chi_q$ of degrees $f_2, \ldots, f_q$ and also with $e = (p-1)/q$ a family of algebraically conjugate "exceptional" characters $\varphi_1, \ldots, \varphi_e$ all of the same degree f_0. For the characters $\chi_1, \ldots, \chi_q$ we have

$$\chi_i(c) = \delta_i = \pm 1 \tag{4.2}$$

and for $\varphi_1, \ldots, \varphi_e$ we have

$$\varphi_j(c) = -\delta_0\eta_j, \quad \delta_0 = \pm 1 \tag{4.3}$$

where η_j is a Gaussian sum, the sum of q pth roots of unity and $\eta_1 + \ldots + \eta_e = -1$. If $q = p - 1$, then $e = 1$ and there is only one exceptional character, which is not distinguishable from the rest. By a classical theorem of Burnside's [9, p. 327], if $q = 1$, G has a normal p-complement so that $q > 1$ for a simple group G. The degrees satisfy congruences modulo p

$$f_i \equiv \delta_i (\text{mod } p),\ i = 1, \ldots, q,\ f_0 \equiv -q\delta_0 (\text{mod } p) \tag{4.4}$$

The degrees have the divisibility property

$$f_i | q(1 + rp) \qquad i = 0, 1, \ldots, q \tag{4.5}$$

Finally the degrees satisfy an equation

$$\delta_1 f_1 + \delta_2 f_2 + \ldots + \delta_q f_q + \delta_0 f_0 = 0 \tag{4.6}$$

For an element $cx = xc$ where $x \in V$ we have

$$\chi_i(cx) = \chi_i(c) = \delta_i,\ i = 1 \ldots q \text{ and } \varphi_j(cx) = \varphi_j(c) = -\delta_0 \eta_j \tag{4.7}$$

For an element y of order relatively prime to p all exceptional characters φ_j have the same value $\varphi(y)$. We have the orthogonality relation

$$\delta_1 \chi_1(y) + \delta_2 \chi_2(y) + \ldots + \delta_q \chi_q(y) + \delta_0 \varphi(y) = 0 \tag{4.8}$$

Taking $y = 1$ gives (4.6) as a special case of (4.8).

As an example of the power of these relations let us show that no finite group contains exactly 21 Sylow 5-subgroups. As 21 does not have any proper divisor of the form $1 + 5k$, it follows from a theorem of the author's [16] that if there is any group with 21 Sylow 5-subgroups, then there is a simple group G with 21 Sylow 5-subgroups. If S_5 is a Sylow 5-subgroup and N_5 its normalizer, then $[G:N_5] = 21$. Let us first consider the possibility that the order of S_5 is 5^r with $r \geqslant 2$. In this case if the representation of G on the 21 cosets of N_5 is not primitive, then G has a representation on 3 or 7 points which cannot be faithful on S_5. Hence G is primitive on 21 points and on S_5 fixes one point and moves the remaining points in four 5-orbits. Thus S_5 is Abelian. By a result of Brodkey [8] it follows that the intersection of all S_5's is the intersection of two of them. But as $21 < 5^2$ it follows that two S_5's intersect in a group of order 5^{r-1}. Thus the intersection of all S_5's, which is a normal subgroup of G, is of order 5^{r-1}. This is in conflict with the simplicity of G. Hence the order of G is divisible by exactly the first power of 5 and the Brauer theory applies.

For $q = 2$ the possible degrees in $B_0(5)$ are $\delta_1 f_1 = 1, \delta_2 f_2 = 6, \delta_0 f_0 = -7$ which we write

$$q = 2 \qquad 1 + 6 - 7 = 0 \tag{4.9}$$

For $q = 4$ the only possible degree equations are

$$1 + 6 - 14 - 14 + 21 = 0$$

$$1 - 4 - 4 - 14 + 21 = 0 \tag{4.10}$$

It has been shown by Stanton [23] that if any degree in the principal block $B_0(p)$ is less than $2p$ then the centralizing group V is the identity for G a simple group. Hence the order g of G is 210 for (4.9) and 420 for (4.10). In the regular representation of a group G of order 210 an involution has 105 transpositions and so is an odd permutation and G has a normal subgroup of index 2. If G is of order 420 then G has 1 or 15 S_7's these being the only divisors of the order of the form $1 + 7k$. If there is only 1 S_7 it is a normal subgroup. If there are 15 S_7's, then N_7 is of order 28 and there is an involution centralizing a 7 element. On 21 points a 7 element has three 7 orbits and an involution centralizing this fixes 7 points of one of the orbits and interchanges the other two so that the involution has 7 transpositions and is an odd permutation. Hence G has a normal subgroup of index 2. We have shown that no simple group G has exactly 21 S_5's and so by the result cited that no group has exactly 21 S_5's.

The relations (4.4), (4.5) and (4.6) lend themselves to a computer attack. Note that the order v of the centralizing group does not enter into these relations. Given the prime p and the number $1 + rp$ of S_p's, there are only a finite number of possibilities for q which is a divisor of $p - 1$ and for the degrees which are divisors of $(p - 1)(1 + rp)$. Such listings were made for primes $p = 5, 7, \ldots, 31$ and appropriate values $1 + rp$ which would make the order of G at most one million.

It is easy to see (using the Brodky theorem) that if a prime p divides g and $p^4 > g$ then p divides g to exactly the first power. For $g < 1{,}000{,}000$, this is true for $p \geqslant 37$. A result of Brauer and Reynolds [5] says that if $p^3 > g$ then G is a known group, either $L_2(p)$ or if p is a Fermat prime $p = 2^m + 1$ then $L_2(2^m)$. For $g < 1{,}000{,}000$ this covers all primes $p > 100$. The primes $p = 37, \ldots, 97$ were treated separately. If we are to obtain any simple group except $L_2(p)$ for one of these there must be a nontrivial factorization

$$(p - 1)(1 + rp) = (1 + up)(vp - 1)$$

with $v > 1$ and $u > 1$. These cases were handled without great difficulty and the main search was made on group orders less than one million divisible only by primes not exceeding 31.

To illustrate the computer contribution to the search for simple groups, let us consider the order $g = 459{,}648 = 2^7 . 3^3 . 7 . 19$. For each prime p greater than 5 the computer lists the possible number $1 + rp$ of S_p's and the order of N_p. Here we have

$$g = 459648 = 2^7 . 3^3 . 7 . 19 \tag{4.11}$$

$$S_7\text{'s } 64 - 7182,\ 36 - 12768,\ 288 - 1596,\ 456 - 1008,$$
$$3648 - 126,\ 2052 - 224,\ 16416 - 28$$
$$S_{19}\text{'s } 1008 - 456$$

Here the only possibility for S_{19}'s is 1008 S_{19}'s with $N_{19} = 456 = 2^3 . 3 . 19$. Thus for an S_{19} we have $q = 2$, $q = 3$ or $q = 6$. For 1008 S_{19}'s and $q = 2, q = 3, q = 6$ our table gives

$q = 2$ no degrees
$q = 3$ no degrees

$$q = 6 \tag{4.12}$$

$\delta_0 f_0$		$\delta_i f_i$		$i = 1, \ldots q$	
	324,	−189,	96,	−56,	−18
32	0	1	2	0	2
32	1	1	0	3	0
−63	0	0	2	2	1
108	0	1	2	2	0
336	0	1	0	2	2

This table lists the number of signed degrees $\delta_i f_i$ in the degree equation (4.6) and omits the principal character $\delta_1 f_1 = 1$ which is always present. The value of $\delta_0 f_0$ is given by itself. Thus the first row of this table corresponds to a degree equation

$$1 - 18 - 18 + 96 + 96 - 189 + 32 = 0 \tag{4.13}$$

Here with $q = 6$, as $|N_{19}| = 456$ we have $v = 4$. By the Stanton condition, this is not possible when there is a degree less than 38 in $B_0(19)$. Hence the only possibility for $B_0(19)$ is given by the fourth row corresponding to the degree equation

$$1 - 56 - 56 + 96 + 96 - 189 + 108 = 0 \tag{4.14}$$

This corresponds to a small piece of the character table for G:

Element x	1	19_1	19_2	19_3
χ_1	1	1	1	1
χ_2	56	-1	-1	-1
χ_3	56	-1	-1	-1
χ_4	96	1	1	1
χ_5	96	1	1	1
χ_6	189	-1	-1	-1
χ_1	108	$-\eta_1$	$-\eta_2$	$-\eta_3$
φ_2	108	$-\eta_2$	$-\eta_3$	$-\eta_1$
φ_3	108	$-\eta_3$	$-\eta_1$	$-\eta_2$

$$\eta_1 + \eta_2 + \eta_3 = -1 \tag{4.15}$$

This shows that there is no element of order 7.19 in G so that for our S_7's we can eliminate the cases of 64, 36, or 288 S_7's. The number of S_7's is therefore 456, 3648, 2052, or 16416. In the last two of these cases we have $|N_7| = 224 = 2^5.7$ and $|N_7| = 28 = 2^2.7$ respectively and so $q = 2$ is the only possibility. But for $q = 2$ in both these cases we find the only possible degree equations to be $1 + 8 - 9 = 0$ and by the Stanton condition we must have $v = 1$ which conflicts with $v = 16$ for $|N_7| = 224$ and $v = 2$ for $|N_7| = 28$. Thus we need only consider 456 or 3648 S_7's. As $v > 1$ in both cases we list only the degrees satisfying the Stanton condition:

$$456\ S_7\text{'s} \qquad N_7 = 1008 = 16.9.7$$

$$q = 2 \text{ no degrees}$$

$$\begin{aligned} q = 3 \qquad & 1 + 57 - 76 + 18 = 0 \\ q = 6 \qquad & 1 - 48 - 48 + 57 + 57 + 57 - 76 = 0 \\ & 1 - 48 - 48 + 57 - 76 - 342 + 456 = 0 \end{aligned} \tag{4.16}$$

For 3648 S_7's we have $N_7 = 126$ and have the same degrees as in (4.16) and also

$$\begin{aligned} q = 6 \qquad & 1 + 36 + 36 + 57 - 76 + 288 - 342 = 0 \\ & 1 + 57 + 57 + 57 - 76 + 288 - 384 = 0 \\ & 1 + 57 - 76 + 288 - 342 - 384 + 456 = 0 \end{aligned} \tag{4.17}$$

If c_7 is an element of order 7 so that $S_7 = \langle c_7 \rangle$ and if ϵ is a primitive 7th root of unity, then if $q = 3$ every character $\psi(c_7)$ of c_7 is a linear combination

$$x.1 + y(\epsilon + \epsilon^2 + \epsilon^4) + z(\epsilon^3 + \epsilon^5 + \epsilon^6)$$
$$= x + y(-1 + \sqrt{-7})/2 + z(-1 - \sqrt{-7})/2$$

where x, y, z are nonnegative integers and $x + 3y + 3z = \psi(1)$. If $q = 6$ then $z = y$ and

$$\psi(c_7) = x \,.\, 1 + y(\epsilon + \epsilon^2 + \epsilon^3 + \epsilon^4 + \epsilon^5 + \epsilon^6) = x - y$$

where $x + 6y = \psi(1)$. One of the orthogonality relations in a character table is $c(x) = |C_G(x)| = \sum_\chi |\chi(x)|^2$ as χ ranges over all irreducible characters. In (4.15) as

$$\varphi_1(c_7) = \varphi_2(c_7) = \varphi_3(c_7) = w$$

is a rational integer we have $w = x - y$ where $x + 6y = 108$ so that $w = 108 - 7y$ and so $|w| \geqslant 3$. Thus $c(c_7) \geqslant 1 + 3^2 + 3^2 + 3^2 = 28$. This excludes the possibility of 3468 S_7's with $q = 6$ and $c(c_7) = 7 \,.\, 3 = 21$. The possibilities for $B_0(7)$ are now restricted to

$$\begin{array}{ll} 456\ S_7\text{'s} & N_7 = 1008 = 16 \,.\, 9 \,.\, 7 \\ q = 3, v = 48 & 1 + 57 - 76 + 18 = 0 \\ q = 6, v = 24 & \\ & \\ \multicolumn{2}{l}{1 - 48 - 48 + 57 + 57 + 57 - 76 = 0} \\ \multicolumn{2}{l}{1 - 48 - 48 + 57 - 76 - 342 + 456 = 0} \\ & \\ 3648\ S_7\text{'s} & N_7 = 126 = 2 \,.\, 9 \,.\, 7 \\ q = 3, v = 6 & 1 + 57 - 76 + 18 = 0 \end{array} \tag{4.18}$$

Any character of an element x_{19} of order 19 is a linear combination of $1, \eta_1, \eta_2, \eta_3$ with nonnegative integral coefficients. Each η is the sum of 6 primitive 19th roots of unity, and η_1, η_2, η_3 are the roots of a rational irreducible cubic equation. Hence in (4.18) the characters χ_{48} of degree 48 are rational for x_{19} and as

$$\chi_{48}(x_{19}) \equiv 48 \pmod{19}$$

we have $|\chi_{48}(x_{19})| \geqslant 9$. But

$$c(x_{19}) = 76 < 9^2 + 9^2$$

a conflict with

$$c(x_{19}) = \sum_\chi |\chi(x_{19})|^2$$

Hence the degrees involving two 48's are impossible. In the degree equation $1 + 57 - 76 + 18 = 0$ by a theorem of the modular theory a χ_{57} and χ_{76} of degrees 57 and 76 are zero for any element whose order is a multiple of 19. It follows that for x_{19} and $x_{19}t$ where t is an involution commuting with x_{19} and for a character of degree 18 we have $\chi_{18}(x_{19}) = -1$ and also $\chi_{18}(tx_{19}) = -1$. This is easily seen to be impossible unless every eigenvalue of t is +1 and $\chi_{18}(t) = 18$ and t is not faithfully represented, contrary to the simplicity of G. In every instance we have reached a conflict and must conclude that there is no simple group of order $459{,}648 = 2^7 . 3^3 . 7 . 19$.

As a further illustration we consider a simple group G of order $g = 979{,}200 = 2^8 . 3^2 . 5^2 . 17$. In the degree equation for $B_0(17)$ at least one further degree f_i besides f_1 must be odd and so a divisor of $3^2 . 5^2 = 225$, and so one of 3, 5, 9, 15, 25, 45, 75, 225. No one of these is congruent to ± 1 modulo 17, and so is not an f_i for $i = 1, \ldots, q$. Thus it must be f_0. We recall that $\delta_0 f_0 \equiv -q \pmod{p}$. As $q = 2, 4, 8, 16$ the possibilities are $f_0 = 9$, $\delta_0 f_0 = 9 \equiv -8 \pmod{17}$, $q = 8$; $f_0 = 15$, $\delta_0 f_0 \equiv -2 \pmod{17}$, $q = 2$; $f_0 = 25$, $\delta_0 f_0 \equiv -8 \pmod{17}$; or $f_0 = 225$, $\delta_0 f_0 \equiv -4 \pmod{17}$. In general from $g = (1 + rp)\, pqv$ we have $g/p \equiv qv \pmod{p}$. In this case with $g = 979{,}200$ and $p = 17$ we have $qv \equiv 4 \pmod{17}$. By the Stanton condition if $f_0 = 9, 15$, or 25 we have $v = 1$ and $q = 2$ or 8 which does not agree with $qv \equiv 4 \pmod{17}$. Hence the only possibility is $f_0 = 225$, $q = 4$, $\delta_0 = -1$. Here since $f_0 \mid 4(1 + 17r)$ we must have $1 + 17r = 225\, m$ with m a divisor of 256. Hence $m = 64$ and there are 14400 S_{17}'s, $q = 4$ and $v = 1$. The possible degree equations for $B_0(17)$ are

$$\begin{aligned} &1 - 16 + 120 + 120 - 225 = 0 \\ &1 - 16 - 16 + 256 - 225 = 0 \\ &1 + 18 - 50 + 256 - 225 = 0 \\ &1 + 256 + 256 - 288 - 225 = 0 \\ &1 - 288 - 288 + 800 - 225 = 0 \end{aligned} \tag{4.19}$$

We can immediately exclude the last of these since the sum of the squares of the degrees is greater than g.

In general in a group G if there are r classes of conjugates $K_1, \ldots, K_r$ then in the group ring $R(G)$ over the complex field the class sums $C_i = \Sigma x$, $x \in K_i$ play a special role. The C_i are a basis for the center of $R(G)$. Here

$$C_i C_j = C_j C_i = \sum_{k=1}^{r} c_{ijk} C_k \tag{4.20}$$

where the c_{ijk} are nonnegative integers. Burnside ([9], p. 316) expressed these coefficients c_{ijk} in terms of the characters

$$c_{ijk} = \frac{g}{c(x_i)c(x_j)} \sum_{m=1}^{r} \frac{\chi_i^m \chi_j^m \overline{\chi_k^m}}{n_m} \tag{4.21}$$

Here χ_i^m is the value of the irreducible character χ^m for an element x_i of the ith class K_i, $n_m = \chi^m(1)$ is the degree of this character and $c(x_i)$, $c(x_j)$ are respectively the orders of the centralizers $C_G(x_i)$, $C_G(x_j)$.

A particular case of (4.21) is that in which $x_i = x_j = \tau$ is an involution transforming x_k into x_k^{-1}. If x_k is of order p then, as shown in Brauer-Fowler [6], Lemma 2A), c_{ijk} is the number of involutions transforming x_k into x_k^{-1}. In the case where $x = x_k$ is of order p and $P = \langle x \rangle$ is its own centralizer and $q | p-1$ is even and $v = 1$ this number is exactly p. As $\chi(x)$ vanishes for characters not in $B_0(p)$ the equation determines $(c(\tau))^2$ in terms of the values of $\chi(\tau)$ and $\chi(x)$ for χ in $B_0(p)$.

Here $N_{17} = \langle a, b \rangle$ with $a^{17} = 1, b^4 = 1, b^{-1}ab = a^4$. N_{17} has order 68 and has the following character table

$h(x)$	1	4	4	4	4	17	17	17	
$c(x)$	68	17	17	17	7	4	4	4	
x	1	a	a^3	a^9	a^{10}	b	b^2	b^3	
λ_0	1	1	1	1	1	1	1	1	
λ_1	1	1	1	1	1	i	-1	$-i$	
λ_2	1	1	1	1	1	-1	1	-1	(4.22)
λ_3	1	1	1	1	1	$-i$	-1	i	
μ_1	4	η_1	η_2	η_3	η_4	0	0	0	
μ_2	4	η_2	η_3	η_4	η_1	0	0	0	
μ_3	4	η_3	η_4	η_1	η_2	0	0	0	
μ_4	4	η_4	η_1	η_2	η_3	0	0	0	

In G the exceptional characters are of degree 225 and we have from the Brauer theory

	1	a	a^3	a^9	a^{10}	
θ_1	225	η_1	η_2	η_3	η_4	
θ_2	225	η_2	η_3	η_4	η_1.	(4.23)
θ_3	225	η_3	η_4	η_1	η_2	
θ_4	225	η_4	η_1	η_2	η_3	

The restriction of θ_1 to N_{17} has the form

$$\theta_1 | N_{17} = \mu_1 + r(\mu_1 + \mu_2 + \mu_3 + \mu_4) + s_0\lambda_0 + s_1\lambda_1 + s_2\lambda_2 + s_3\lambda_3 \qquad (4.24)$$

with nonnegative integers r, s_0, s_1, s_2, s_3. Writing $s = s_0 + s_1 + s_2 + s_3$ the facts that $\theta_1(1) = 225$ and $\theta_1(a) = \eta_1$ give $225 = 4 + 16r + s$, $\eta_1 = \eta_1 - r + s$, and so $r = s = 13$. From (4.22) it now follows that $-13 \leqslant \theta_1(b^2) \leqslant 13$. As the eigenvalues of $\theta_1(b^2)$ are all +1's or −1's and as its determinant is +1, the number of eigenvalues which are −1 is an even number and it follows that $\theta_1(b^2) \equiv \theta_1(1) \pmod 4$. Hence

$$\theta_1(b^2) = 13, 9, 5, 1, -3, -7, -11 \qquad (4.25)$$

From the modular theory, since $256 = 2^8$ is the highest power of 2 dividing g, then, if $\chi(1) = 256m$, $\chi(x) = 0$ for any element x whose order is even. Thus if $\chi(1) = 256$, $\chi(b^2) = \chi(b) = \chi(b^3) = 0$. With this fact and calculations similar to that used to derive (4.25) we find for the degrees of the first 4 equations of (4.19)

$\chi(1)$	$\chi(b^2)$	
16	0	
18	2, −2	
50	2, −2	(4.26)
120	8, 4, 0, −4, −8	
225	13, 9, 5, 1, −3, −7, −11	
256	0	
288	16, 12, 8, 4, 0, −4, −8, −12, −16	

And of course we also have the orthogonality relation $\sum_{\chi} \chi(a)\chi(b^2) = 0$.

For the degrees $1 - 16 + 120 + 120 - 225 = 0$ we have the partial character table

	1	a	b^2
χ_1	1	1	1
χ_2	16	−1	0
χ_3	120	1	u
χ_4	120	1	v
θ_1	225	η_1	w
θ_2	225	η_2	w
θ_3	225	η_3	w
θ_4	225	η_4	w

(4.27)

where by orthogonality $1 + u + v - w = 0$. Also u, v, w take only the values listed in (4.26). In (4.21) with $x_i = x_j = b^2 = \tau$ and $x_k = a$ by Brauer-Fowler $c_{ijk} = 17$ and we have

$$17 = \frac{g}{(c(\tau))^2}\left(1 + \frac{u^2}{120} + \frac{v^2}{120} - \frac{w^2}{225}\right) \tag{4.28}$$

which simplifies, using $w = 1 + u + v$, to

$$(c(\tau))^2 = 32(1792 + 7u^2 - 16uv + 7v^2 - 16u - 16v) \tag{4.29}$$

None of the permissible values 8, 4, 0, −4, −8 for u and v make the right hand side of (4.29) a square. Hence the degrees $1 - 16 + 120 + 120 - 225 = 0$ are to be excluded. Similar, but easier calculations exclude the degrees $1 - 16 - 16 +$

256 − 225 = 0 and 1 + 256 + 256 − 288 − 225 = 0. In the only remaining case we have

order	1	17	17	17	17	2	
$c(x)$	g	17	17	17	17	256	
$h(x)$	1	57600	57600	57600	57600	3825	
x	1	a	a_3	a^9	a^{10}	b^2	
ρ_0	1	1	1	1	1	1	
ρ_1	18	1	1	1	1	±2	
ρ_2	50	−1	−1	−1	−1	±2	(4.30)
ρ_3	256	1	1	1	1	0	
θ_1	225	η_1	η_2	η_3	η_4	1	
θ_2	225	η_2	η_3	η_4	η_1	1	
θ_3	225	η_3	η_4	η_1	η_2	1	
θ_4	225	η_4	η_1	η_2	η_3	1	

The character ρ_1 of degree 18 is rational. As there are 20 primitive 25 roots of unity and $20 > 18$ there is no element of order 25 in G and so in G a Sylow 5-group S_5 is elementary Abelian. Since 25 divides both 50 and 225, the character ρ_2 and the θ's are zero for elements whose orders are multiples of 5.

For the rational character ρ_1 of degree 18 a 5-element has character $r\,.\,1 + s(\epsilon + \epsilon^2 + \epsilon^3 + \epsilon^4) = r - s$ where ϵ is a primitive 5th root of unity and $r + 4s = 18$. Hence for 5-elements x_5 the possible characters in $B_0(17)$ are

	1	a	b^2	x_5	x_5	x_5	x_5	
ρ_0	1	1	1	1	1	1	1	
ρ_1	18	1	±2	13	8	3	−2	
ρ_2	50	−1	±2	0	0	0	0	(4.31)
ρ_3	256	1	0	−14	−9	−4	1	
θ	225	η	1	0	0	0	0	

For any element x, where $h(x)$ is the number of elements in the conjugate class of x and if χ is any irreducible character it is well known that $h(x)\chi(x)/\chi(1)$ is an algebraic integer. Hence above $h(x)\rho_1(x)/18$ and $h(x)\rho_3(x)/256$ are rational integers. Also for an x_5, $17|h(x)$ since then S_{17} is self centralizing. Hence when $\rho_1(x_5) = 13$, $\rho_3(x_5) = -14$, $128\,.\,9\,.\,17|h(x_5)$ and so $c(x_5)|2.25 = 50$ while $1 + (13)^2 + (-14)^2 > 50$ a conflict. Also when $\rho_1(x_5) = 8$, $\rho_3(x_5) = -9$, $256\,.\,9\,.\,17|h(x_5)$ and $c(x_5)|25$ while $1 + 8^2 + (-9)^2 > 25$ a conflict. When $\rho_1(x_5) = 3$, $\rho_3(x_5) = -4$, $3\,.\,64\,.\,17|h(x_5)$ and $c(x_5)|3\,.\,4\,.\,25 = 300$. When $\rho_1(x_5) = -2$, $\rho_3(x_5) = 1$ then $256\,.\,9\,.\,17|h(x_5)$ and $c(x_5)|25$. These last two are the possible characters for 5-elements. Now let us consider ρ_1 restricted to an S_5 of order 25 and suppose there are m elements x for which $\rho_1(x) = -2$ and n for which $\rho_1(x) = 3$, and let σ_0 be the principal character of S_5

$$\begin{array}{cccc} & 1 & m & n \\ \sigma_0 & 1 & 1 & 1 \\ \rho_1|S_5 & 18 & -2 & 3 \end{array} \tag{4.32}$$

Here $m + n = 24$ and for any 5-element x $\rho_1(x^i) = \rho_1(x), i = 1, 2, 3, 4$, so that m and n are both multiples of 4. Then if t is the multiplicity of σ_0 in ρ_1 restricted to S_5 we have by the usual orthogonality relations

$$18 - 2m + 3n = 25t, \qquad m + n = 24 \tag{4.33}$$

Here with m, n, t nonnegative integers and m and n multiples of 4 the only solution is $m = 8, n = 16, t = 2$. As we have noted if x_0 is a 5-element with $\rho_1(x_0) = -2$ then $c(x_0) = 25$ and $h(x_0) = 39168$. Each S_5 contains exactly 8 of these elements and each of these elements is in exactly one S_5. Hence if there are k such classes the number of S_5's is $k.(39168)/8 = 4896k = 2^5.3^2.17k$. Hence k divides 8 and $k \equiv 1 \pmod 5$ so that $k = 1$ and so there is exactly one class of 5-elements with $\rho_1(x_0) = -2$ and there are 4896 S_5's. A 5-element in more than one S_5 will be centralized by the group which they generate. As such a centralizer is of order at most 300, it is easy to see that such an $\langle x_i \rangle$ is in 6 S_5's and $C_G(x_i) = \langle x_i \rangle \times A_5$ is of order 300. As $4896 = 1 + 5 + 5 + 5 + 5 + 25M$ it follows that an S_5 has 4 of its subgroups of order 5 contained in other S_5's. Hence each 5-element x_i for which $\rho_1(x_i) = 3$ is contained in 6 S_5's and has $c(x_i) = 300$. It now follows from a simple count that there are exactly 4 such classes.

An involution v centralizing one of these 5-elements x_i cannot be conjugate to b^2 whose centralizer is of order 256. Hence the product of v with a conjugate of itself or with a conjugate of b^2 cannot be a 17-element since in the first instance it would normalize a 17-element and so be conjugate to b^2, and in the second instance two involutions whose product is of odd order are conjugate. Writing $\rho_1(v) = x$, $\rho_2(v) = y$, $\rho_3(v) = 0$, $\theta_i(v) = z$ we have equations

$$\begin{aligned} &1 + x - y - z = 0 \\ &1 + \frac{x\rho_1(b^2)}{18} - \frac{y\rho_2(b^2)}{50} - \frac{z\theta(b^2)}{225} = 0 \\ &1 + \frac{x^2}{18} - \frac{y^2}{50} - \frac{z^2}{225} = 0 \end{aligned} \tag{4.34}$$

There is no solution of these in integers when $\rho_1(b^2) = \rho_2(b^2) = -2$, $\theta(b^2) = 1$ while for $\rho_1(b^2) = \rho_2(b^2) = 2$, $\theta(b^2) = 1$ we find $x = -6$ and $x = -21$ as the possible values for x. Since -21 is not possible as a character for an element in a representation of degree 18, we must have $\rho_1(v) = -6$ and from (4.32) $y = \rho_2(v) = 10$, $z = \theta(v) = -15$. From θ_1, $h(v)(-15)/225$ is an integer so that $15|h(v)$. Also $17|h(v)$ so that $15 . 17 = 255$ divides $h(v)$.

From the orthogonality relations $\sum_{x \in G} |\theta_1(x)|^2 = g = 979{,}200$ and $\sum_{x \in G} \theta_1(x) = 0$.

Excluding the known values of x, as the identity, a 17-element, or a conjugate of b^2 we have for the remaining x's, whose characters are rational integers

$$\Sigma'\theta_1(x)^2 = 175950$$
$$\Sigma'\theta_1(x) = 53550 \tag{4.35}$$

For an element v, as $\theta_1(v) = -15$ and $h(v) = 255m$ we have $255m(-15)^2 \leqslant 175{,}950$ or $57375m \leqslant 175{,}950$ so that $m = 1, 2$, or 3. The value $m = 3$ would force $\Sigma'\theta_1(x)^2 = 3825$ and $\Sigma'\theta_1(x) = 65025$ for the remaining elements, an obvious absurdity for rational integers. Hence m is 1 or 2, and $|C_G(v)|$ has order 3840 or 1920.

As the centralizer of a 5-element x_i is $\langle x_i \rangle \times A_5$ there are 15 conjugate centralizing involutions. As there are $4 . 3264 = 13056$ 5-elements of this kind there are $15 . 13056 = 195{,}840$ pairs (x, v) of 5-elements x and centralizing involutions v. As $C_G(v) = H$ is of order 1920 or 3840 and has v in its center, H contains 1, 16, or 96 Sylow 5-subgroups and so 4, 64, or 384 5-elements. If the number of such involutions v was only 255 this would yield at most $255 . 384 = 97920$ pairs (x, v) of 5-elements and commuting involutions which is only half of the 195,840 there are. Hence the number of such involutions v is $2 . 255 = 510$. Initially it is not clear whether these consist of a single class of 510 elements or two classes with 255 in each. For remaining elements x we have

$$\Sigma'\theta_1(x)^2 = 61200$$
$$\Sigma'\theta_1(x) = 61200 \tag{4.36}$$

and so there remain 61200 elements for which $\theta_1(x) = 1$, and for all others $\theta_1(x) = 0$.

Completion of the character table for $B_0(7)$ is now reasonably straightforward. The 61200 elements for which $\theta_1(x) = 1$ are necessarily 2-elements. Each of the four classes $\{x_i\}$ of order 5 whose centralizer is of order 300, being $\langle x_i \rangle \times A_5$, gives a class $\{r_i\}$ of order 15 with $c(r_i) = 15$, $h(r_i) = 65280$, and a class $\{s_i\}$ of order 10 with $c(s_i) = 20$, $h(s_i) = 48960$. There remain 174,080 elements whose orders are multiples of 3, and for all these $\rho_1(x) = 0$, $\theta_i(x) = 0$. If the order of the element is even then also $\rho_3(x) = 0$. For a 3-element x, $\rho(x) \equiv \rho(1) \pmod 3$.

The only possible characters for these elements are

	1	a	u_1	u_2	u_3	k
ρ_0	1	1	1	1	1	1
ρ_1	18	1	0	0	0	0
ρ_2	50	−1	5	2	−1	1
ρ_3	256	1	4	1	−2	0
θ_1	225	η_1	0	0	0	0

(4.37)

Other values are easily seen to be impossible. For $h(u)\rho_2(u)/50$ and $h(u)\rho_3(u)/256$ are integers. Thus if $\rho_2(u) = 8$ and $\rho_3(u) = 7$ it follows that $25 . 256 . 17$ divides $h(u)$ so that $c(u)$ divides 9, and as $1^2 + 8^2 + 7^2 > 9$ this is a conflict. For the elements in (4.37) we have $c(u_1)|180, c(u_2)|9, c(u_3)|18$. An element k is of order a multiple of 6. Writing $c(v) = 1920s$, $s = 1$ or 2, the coefficient $c_{kva}(= c_{ijk}$ with $x_i = k, x_j = v, x_k = a)$ is given by

$$c_{kva} = \frac{256 . 9 . 25 . 17}{c(k)c(v)}\left(1 - \frac{10}{50}\right)$$
$$= \frac{8 . 3 . 17}{c(k)s} \qquad (4.38)$$

Thus $c(k)|24$ and if $s = 2$, $c(k)|12$. Hence we have $5440m_1$, $108{,}800m_2$, $54400m_3$ elements whose characters are of type u_1, u_2, u_3 respectively and $40800m_4$ of type k. Here $m_1 \geqslant 1$ since there is an element of order 3 in the centralizer of a 5-element and $m_4 \geqslant 1$ as there is an element of order 6 in $C_G(v)$. From the orthogonality $(\rho_2, \rho_3) = 0$, since ρ_3 vanishes on the 61200 2-elements, we have

$$-217600 + 20(5440m_1) + 2(108800m_2) + 2(54400m_3) = 0 \qquad (4.39)$$

Since $m_1 \geqslant 1$ this is possible only if $m_1 = 1, m_2 = 0, m_3 = 1$ or $m_1 = 2, m_2 = 0, m_3 = 0$. Counting the 174,080 elements

$$5440m_1 + 108800m_2 + 54400m_3 + 40800m_4 = 174080 \qquad (4.40)$$

$m_1 = 1, m_2 = 0, m_3 = 1$ does not make m_4 an integer so that the only possible solution is $m_1 = 2, m_2 = 0, m_3 = 0, m_4 = 4$.

There is either one class of order 3 whose centralizer is of order 90 or two 3-classes with centralizer of order 180 for each. But a group of order 90 has a subgroup of order 45 and in this an S_5 is normal and so central. But no 5-element has a centralizer of order divisible by 9. By the same token an element of order 9 cannot have a centralizer of order 180. Hence there are two classes of elements of order 2 which we now call $\{u_1\}$ and $\{u_2\}$ and each of these has $c(u) = 180$. As an element of order 3 is conjugate to its inverse in an A_5 one of the classes $\{u_1\}$, $\{u_2\}$ is conjugate to its inverse and so both. In $C_G(v)$ there is only one conjugate class of 3-elements (and inverses) and so if there was only one class of involutions both $\{u_1\}$ and $\{u_2\}$ would be the same class. Hence there are two classes $\{v_1\}$ and $\{v_2\}$ and $c(v_1) = c(v_2) = 3840$ and v_1 is centralized by one of the classes, say u_1 and v_2 by u_2. Then $u_1 v_1 = k_1$ and $u_2 v_2 = k_2$ are nonconjugate classes whose centralizer has order dividing 12 from (4.38) giving two classes $\{k_1\}$ $\{k_2\}$ each with a multiple of 81600 elements and so exactly this value.

The characters for $B_0(17)$ are now almost fully determined:

order	1	17	2	5	5	15	10	3		6		2		2^j
$c(x)$	g	17	256	25	300	15	20	180		12		3840		
$h(x)$	1	57600	3825	39168	3264	65280	48960	5440		81600		255		
x	1	a	b^2	x_0	x_i	r_i	s_i	u_1	u_2	k_1	k_2	v_1	v_2	
ρ_0	1	1	1	1	1	1	1	1	1	1	1	1	1	1
ρ_1	18	1	2	−2	3	0	−1	0	0	0	0	−6	−6	
ρ_2	50	−1	2	0	0	0	0	5	5	1	1	10	10	
ρ_3	256	1	0	1	−4	−1	0	4	4	0	0	0	0	0
θ_1	225	η_1	1	0	0	0	0	0	0	0	0	−15	−15	1

(4.41)

Here for x_i, r_i, s_i there are 4 classes, $i = 1, 2, 3, 4$. We must still investigate the characters of the 61200 2-elements which remain which are of type w

	1	a	w
ρ_0	1	1	1
ρ_1	18	1	s
ρ_2	50	−1	s
ρ_3	256	1	0
θ_1	225	η_1	1

(4.42)

Here s is an even integer and we find for coefficients in the class multiplication table

$$c_{wva} = 17(16 - 8s)/c(w)$$
$$c_{wx_1a} = 32\,.\,17(6 + s)/c(w) \tag{4.43}$$

Hence $-6 \leqslant s \leqslant 2$ and the possible types are

$h(x)$		$15300m_1$	$61200m_2$	$30600m_3$	$61200m_4$	$3825m_5$
x	1	w_1	w_2	w_3	w_4	w_5
ρ_0	1	1	1	1	1	1
ρ_1	18	−6	−4	−2	0	2
ρ_2	50	−6	−4	−2	0	2
ρ_3	256	0	0	0	0	0
θ_1	225	1	1	1	1	1

(4.44)

where we have used (4.43) to bound the centralizers of the elements.

$\sum_x |\rho_1(x)|^2 = g$ gives

$$550800m_1 + 979200m_2 + 122400m_3 + 15300m_5 = 244800 \qquad (4.45)$$

Hence $m_1 = 0$, $m_2 = 0$ and $8m_3 + m_5 = 16$. $(\rho_1, \rho_0) = 0$ gives, as $m_1 = 0$, $m_2 = 0$,

$$-61200m_3 + 7650m_5 = 0 \qquad (4.46)$$

or

$$m_5 = 8m_3$$

Thus $m_3 = 1$ and $m_5 = 8$ and as there are only 61200 elements in all necessarily $m_4 = 0$.

$c(x)$	g	32	32
$h(x)$	g	30600	$3825m_5 = 30600$
x	1	y	w
ρ_0	1	1	1
ρ_1	18	−2	2
ρ_2	50	−2	2
ρ_3	256	0	0
θ_1	225	1	1

(4.47)

This completely determines the characters in $B_0(17)$ except for the ambiguity that the class labelled w may in fact be as many as 8 classes with the same values in $B_0(17)$. This ambiguity is only removed later.

The skew-symmetric tensor product of ρ_1 with itself has the character

$$\varphi(x) = (\rho_1(x)^2 - \rho_1(x^2))/2 \qquad (4.48)$$

Calculation of its norm showed this to be an irreducible character which is called φ_1 in Table 4.51. The symmetric tensor product of ρ_1 with itself has character $\psi(x)$:

$$\psi(x) = (\rho_1(x)^2 + \rho_1(x^2))/2 \qquad (4.49)$$

Here $\psi(x)$ contains ρ_0 once and from its norm contains two further irreducibles so that

$$\psi = \rho_0 + \varphi_2 + \varphi_3 \qquad (4.50)$$

Here ψ does not contain any further character in $B_0(17)$. We have the problem of constructing φ_2 and φ_3 separately given their sum $\varphi_2 + \varphi_3$. This involves difficulties but not insuperable ones. Taking further products, norms, and inner products the complete character table (4.51) was constructed.

The writer has shown elsewhere [15] that given this table for a simple group of order 979,200, it necessarily contains a subgroup H of order 7200 which is the direct product of two A_5's extended by an involution which interchanges them. $[G:H] = 136$ and H has orbits of lengths 1, 60, 75. Constructing the permutations generating H it was shown possible to extend H by a further permutation to give a transitive permutation group G of order 979,200 in an essentially unique way. Hence the simple group of order 979,200 is unique and so is the known symplectic group $Sp_4(4)$ of this order.

5. Permutation and matrix representations of groups

No single method of presentation seems to be best for all groups. For Abelian groups, generators and relations seem to be the best presentation, and if the group is finitely generated the generators and relations can be put into a canonical form and most questions about the group can be readily answered. For non-Abelian groups, as the word problem is unsolvable in general for finitely presented groups, almost no question can be answered directly in terms of the relations. If the group is finite, coset enumeration will put the group into a permutation form.

The writer has a computer program, written by Richard Lane, which will multiply permutations, find inverses, and also find coset representatives for the stabilizer of a point. One particular order, which seemed somewhat fanciful at first, has proved to be very useful. This is the order to take a given set of generating permutations and to calculate up to, say 200, random products of these generators, recording the generating word in each case, but not including the same permutation twice. In this way it is possible to find certain things in practise for which no straightforward process is known to the writer. For example we may be given several generators which generate a group transitive on 127 points. By the Sylow theorems we know that they must generate an element which is a cycle of length 127, but random outputs in cycle form are in practise the easiest way to find such a thing. Outputs can be in cycle form or in two-rowed permutation form. Each of these forms has its own virtues.

Another program written by Jonathan Hall finds solutions for equations in permutations by systematic trials. This is useful in finding permutation extensions of a known permutation group. Thus if we wish to find a permutation group G on n points and have the stabilizer G, of the point 1, then an involution t in G interchanging 1 and 2 will induce an automorphism in G_{12} by conjugation. Also if x is an element in G_1 moving 2, then there may be elements y and z in G_1 such that $txt = ytz$. If y and z can be determined, the conditions $tG_{12}t = G_{12}$ and one or more equations of the form $txt = ytz$ will determine t.

If a permutation group G has been constructed by coset enumeration then the stabilizer of a point G_1 is given by the construction. But if the permutation group

TABLE 4.51 $Sp_4(4)$ $g = 979{,}200 = 256 . 9 . 25 . 17$

Order				17		2	5			5				15				10		3		6		2			4
$c(x)$				17		256	25			300				15				20		180		12		3840			32
$h(x)$				57600		3825	39168			3264				65280				48960		5440		81600		255			30600
x	1	a	a^3	a^9	a^{10}	b^2	x_0	x_1	x_1^2	x_2	x_2^2	r_1	r_1^2	r_2	r_2^2	s_1	s_1^7	s_2	s_2^7	u_1	u_2	k_1	k_2	v_1	v_2	w	y
ρ_0	1	1	1	1	1	1	1	1	1	1	1	1	1	1	1	1	1	1	1	1	1	1	1	1	1	1	1
ρ_1	18	1	1	1	1	2	−2	3	3	3	3	0	0	0	0	−1	−1	−1	−1	0	0	0	0	−6	−6	2	−2
ρ_2	50	−1	−1	−1	−1	2	0	0	0	0	0	0	0	0	0	0	0	0	0	5	5	1	1	10	10	2	−2
ρ_3	256	1	1	1	1	0	1	−4	−4	−4	−4	−1	−1	−1	−1	0	0	0	0	4	4	0	0	0	0	0	0
θ_1	225	η_1	η_2	η_3	η_4	1	0	0	0	0	0	0	0	0	0	0	0	0	0	0	0	0	0	−15	−15	1	1
θ_2	225	η_2	η_3	η_4	η_1	1	0	0	0	0	0	0	0	0	0	0	0	0	0	0	0	0	0	−15	−15	1	1
θ_3	225	η_3	η_4	η_1	η_2	1	0	0	0	0	0	0	0	0	0	0	0	0	0	0	0	0	0	−15	−15	1	1
θ_4	225	η_4	η_1	η_2	η_3	1	0	0	0	0	0	0	0	0	0	0	0	0	0	0	0	0	0	−15	−15	1	1
φ_1	153	0	0	0	0	−7	3	3	3	3	3	0	0	0	0	−1	−1	−1	−1	0	0	0	0	9	9	1	1
φ_2	85	0	0	0	0	5	0	5	5	0	0	−1	−1	0	0	1	1	0	0	4	−5	0	−1	21	5	1	1
φ_3	85	0	0	0	0	5	0	0	0	5	5	0	0	−1	−1	0	0	1	1	−5	4	−1	0	5	21	1	1
φ_4	34	0	0	0	0	2	−1	4	4	−1	−1	1	1	−1	−1	0	0	−1	−1	1	4	1	0	10	−6	−2	2
φ_5	34	0	0	0	0	2	−1	−1	−1	4	4	−1	−1	1	1	−1	−1	0	0	4	1	0	1	−6	10	−2	2
φ_6	340	0	0	0	0	4	0	−5	−5	0	0	1	1	0	0	−1	−1	0	0	1	−5	1	−1	4	20	0	0
φ_7	340	0	0	0	0	4	0	0	0	−5	−5	0	0	1	1	0	0	−1	−1	−5	1	−1	1	20	4	0	0
φ_8	204	0	0	0	0	−4	−1	D	E	$3A$	$3B$	A	B	0	0	1	1	$-A$	$-B$	3	0	−1	0	−4	12	0	0
φ_9	204	0	0	0	0	−4	−1	E	D	$3B$	$3A$	B	A	0	0	1	1	$-B$	$-A$	3	0	−1	0	−4	12	0	0
φ_{10}	204	0	0	0	0	−4	−1	$3B$	$3A$	D	E	0	0	A	B	$-B$	$-A$	1	1	0	3	0	−1	12	−4	0	0
φ_{11}	204	0	0	0	0	−4	−1	$3A$	$3B$	E	D	0	0	B	A	$-A$	$-B$	1	1	0	3	0	−1	12	−4	0	0
φ_{12}	255	0	0	0	0	−1	0	$5A$	$5B$	0	0	$-A$	$-B$	0	0	A	B	0	0	−3	0	1	0	−17	15	−1	−1
φ_{13}	255	0	0	0	0	−1	0	$5B$	$5A$	0	0	$-B$	$-A$	0	0	B	A	0	0	−3	0	1	0	−17	15	−1	−1
φ_{14}	255	0	0	0	0	−1	0	0	0	$5A$	$5B$	0	0	$-A$	$-B$	0	0	A	B	0	−3	0	1	15	−17	−1	−1
φ_{15}	255	0	0	0	0	−1	0	0	0	$5B$	$5A$	0	0	$-B$	$-A$	0	0	B	A	0	−3	1	1	15	−17	−1	−1
φ_{16}	51	0	0	0	0	3	1	$3+A$	$3+B$	$-3A$	$-3B$	A	B	0	0	$-B$	$-A$	A	B	3	0	−1	0	−13	3	−1	−1
φ_{17}	51	0	0	0	0	3	1	$3+B$	$3+A$	$-3B$	$-3A$	B	A	0	0	$-A$	$-B$	B	A	3	0	−1	0	−13	3	−1	−1
φ_{18}	51	0	0	0	0	3	1	$-3B$	$-3A$	$3+A$	$3+B$	0	0	A	B	B	A	$-B$	$-A$	0	3	0	−1	3	−13	−1	−1
φ_{19}	51	0	0	0	0	3	1	$-3A$	$-3B$	$3+B$	$3+A$	0	0	B	A	A	B	$-A$	$-B$	0	3	0	−1	3	−13	−1	−1

$r_1 = x_1 u_1$, $s_1 = x_1 v_1$, $k_1 = u_1 v_1$, η_i = Gauss sums of 17th roots of unity

$r_2 = x_2 u_2$, $s_2 = x_2 v_2$, $k_2 = u_2 v_2$

$A = (1+\sqrt{5})/2$, $B = (1-\sqrt{5})/2$, $D = -3+4A$, $E = -3+4B$

is given in some other way, for example by transitive extension of a stabilizer G_1, it can be more difficult to determine the exact stabilizer since it is clearly easy to find elements extending G_1 to a transitive group G^* whose stabilizer G_1^* is a group larger than G_1. In general writing $G_1 = H$

$$G = H + Hx_2 + \ldots + Hx_n, \quad x_1 = 1 \tag{5.1}$$

Then if $\langle a_1, \ldots, a_r \rangle = G$, generators for H will be u_{ij} $i = 1, \ldots, n, j = 1, \ldots, r$

$$u_{ij} = x_i a_j \varphi(x_i a_j)^{-1} \tag{5.2}$$

where $\varphi(x_i a_j) = x_k$ if $x_i a_j \in Hx_k$, and in the case of a permutation group on 1, $\ldots, n$ with $H = G_1$ $\quad x_i$ is an element $\begin{pmatrix} 1 & \ldots \\ i & \ldots \end{pmatrix}$. The Hall-Janko simple group G of order 604,800 was constructed as a transitive extension of $U_3(3)$ with orbit lengths 1, 36, 63. Here $U_3(3) = \langle a, b \rangle$ with $a^7 = 1$ and $b^8 = 1$ and $G = \langle a, b, t \rangle$ where t was an involution with $tat = a^{-1}$. Peter Swinnerton-Dyer verified on the Titan computer at Cambridge that the 300 elements $rx_i\varphi(rx_i)^{-1}$, $r = a, b, c, i = 1, \ldots, 100$ were in fact elements of $U_3(3)$ proving that $\langle a, b, c \rangle$ was in fact of order 604,800. From this the simplicity of the group was immediate. There is another way of regarding this permutation group. It is a rank 3 group, which is to say that the stabilizer of a point b has three orbits (b), $\Delta(b)$, $\Gamma(b)$ and in this case we always have $|\Delta(b)| = 36$, $|\Gamma(b)| = 63$. We may define a graph $\mathcal{G}$ by joining each point b to the points of $\Delta(b)$. This is a strongly regular graph in that every point is joined to 36 points. Two points which are joined are both joined to exactly 14 others and two points which are not joined to each other are both joined to exactly 12 other points. It can be verified that $\langle a, b, c \rangle$ is an automorphism of group of the graph $\mathcal{G}$. The full automorphism group of the graph is the Hall-Janko group extended by an automorphism of order 2.

The theory of rank 3 groups G was developed by D. G. Higman [17]. Let G be a primitive rank 3 group of even order, so that G is transitive on n points and let the orbits of a stabilizer G_a be

$$\{a\}, \quad \Delta(a), \quad \Gamma(a) \tag{5.3}$$

of lengths 1, k, l respectively with $1 + k + l = n$, $k \leqslant l$. The orbit intersections are given by

$$|\Delta(a) \cap \Delta(b)| = \begin{cases} \lambda \text{ for } b \in \Delta(a) \\ \mu \text{ for } b \in \Gamma(a) \end{cases} \tag{5.4}$$

Associated with the orbits Δ is an n by n incidence matrix A

$$A = [a_{ij}], i, j = 1, \ldots, n$$

$$a_{ij} = \begin{cases} 1 \text{ if } j \in \Delta(i) \\ 0 \text{ if } j \notin \Delta(i) \end{cases} \tag{5.5}$$

Let J be the n by n matrix in which every entry is 1, and let I be the identity matrix. Define B as $B = J - I - A$. Then A^T being the transpose of A

$$AA^T = kI + \lambda A + \mu B$$

$$A^T = A \tag{5.6}$$

$$(A - kI)(A^2 - (\lambda - \mu)A - (k - \mu)I) = 0$$

The parameters k, l, λ, μ satisfy various relations. One eigenvalue of A is k (with multiplicity one because of the primitivity of G). The other two eigenvalues are s, t given by

$$\begin{Bmatrix} s \\ t \end{Bmatrix} = ((\lambda - \mu) \pm \sqrt{d})/2, \qquad d = (\lambda - \mu)^2 + 4(k - \mu) \tag{5.7}$$

The multiplicities f_2 and f_3 of these eigenvalues are given by

$$\begin{Bmatrix} f_2 \\ f_3 \end{Bmatrix} = \frac{2k + (\lambda - \mu)(k + l) \mp \sqrt{d}(k + l)}{\mp 2\sqrt{d}} \tag{5.8}$$

We can obtain various conclusions which we state as a lemma:

Lemma. *Either* (*a*) $k = 1, \mu = \lambda + 1 = k/2$ *and* $f_2 = f_3 = k$ *or* (*b*) $d = (\lambda - \mu)^2 + 4(k -$ *is a square, and* f_2, f_3 *in (5.8) are rational integers.*

These properties hold for a strongly regular graph, namely a graph on n points in which every point is joined to exactly k others, two points which are joined to each other are both joined to exactly λ other points, and two points which are not joined are both joined to exactly μ others. The condition $\mu > 0$ corresponds on the one hand to primitivity of and on the other to connectivity of the graph. For the rank 3 group $1, f_2, f_3$ are the degrees of the irreducible representations into which the permutation representation decomposes. Tabulations of parameters satisfying the conditions have been made by Richard Lane.

The construction of the Hall-Janko group as a rank 3 permutation group in August 1967 stimulated a search for sporadic simple groups among rank 3 groups. D. G. Higman and C. Sims [18] constructed the simple group of order 44,352,000 as a rank 3 extension on 100 points of the Mathieu group M_{22} on orbits of length 1, 22, 77. This was done in September 1967. M. Suzuki ([7], 113–120) showed that $G_2(4)$ is a rank 3 extension of HJ (the Hall-Janko group) of degree 416 with orbits 1, 100, 315, and in turn that a new simple group on 1782 points with stabilizer $G_2(4)$ and orbits 1, 416, 1365 $\lambda = 100$ and $\mu = 96$ could be constructed, the group being of order 448,345,497,600. Similarly McLaughlin ([7], 109–112) in January 1968 found a rank 3 extension of $U_4(3)$ on 275 points with orbits 1, 112, 162 the group being of order 898,128,000.

TABLE 5.9

α

	1	7	13	19
1	$-i$	1	-1	-1
7	i	-1	-1	-1
13	1	$-i$	i	$-i$
19	i	1	1	-1

$\frac{1}{2}$

	2	8	14	20
2	-1	-1	-1	i
8	$-i$	i	$-i$	1
14	1	1	-1	i
20	1	-1	-1	$-i$

$\frac{1}{2}$

	3	9	15	21
3	$-i$	1	-1	-1
9	i	-1	-1	-1
15	1	$-i$	i	$-i$
21	i	1	1	-1

	4	10	16	22
4	1	0	0	0
10	0	1	0	0
16	0	0	1	0
22	0	0	0	1

$\frac{1}{2}$

	5	11	17	23
5	-1	i	1	1
11	-1	$-i$	1	-1
17	-1	i	-1	-1
23	$-i$	1	$-i$	i

$\frac{1}{2}$

	6	12	18	24
6	-1	-1	-1	i
12	$-i$	i	$-i$	1
18	1	1	-1	i
24	1	-1	-1	$-i$

	25	26	27	28
25	1	0	0	0
26	0	1	0	0
27	0	0	1	0
28	0	0	0	1

All other entries in α are zero.

β

$$\begin{matrix} x_1 x_2 x_3 x_4 x_5 x_6 x_7 x_8 x_9 x_{10} x_{11} x_{12} x_{13} x_{14} x_{15} x_{16} x_{17} x_{18} x_{19} x_{20} x_{21} x_{22} x_{23} x_{24} \\ x_2 x_3 x_4 x_5 x_6 x_7 x_8 x_9 x_{10} x_{11} x_{12} x_{13} x_{14} x_{15} x_{16} x_{17} x_{18} x_{19} x_{20} x_{21} x_{22} x_{23} x_{24} x_1 \end{matrix}$$

β: $\frac{1}{2}$

	x_{25}	x_{26}	x_{27}	x_{28}
x_{25}	$-i$	$-i$	i	-1
x_{26}	i	$-i$	$-i$	-1
x_{27}	$-i$	i	$-i$	-1
x_{28}	1	1	1	i

$-\delta$

$$\begin{matrix} x_1 x_2 x_3 x_4 x_5 x_6 x_7 x_8 x_9 x_{10} x_{11} x_{12} x_{13} x_{14} x_{15} x_{16} x_{17} x_{18} x_{19} x_{20} x_{21} x_{22} x_{23} x_{24} \\ x_1 x_6 x_{11} x_{16} x_{21} x_2 x_7 x_{12} x_{17} x_{22} x_3 x_8 x_{13} x_{18} x_{23} x_4 x_9 x_{14} x_{19} x_{24} x_5 x_{10} x_{15} x_{20} \end{matrix}$$

$-\delta$ $\frac{1}{2}$

	x_{25}	x_{26}	x_{27}	x_{28}
x_{25}	-1	1	1	$-i$
x_{26}	1	1	-1	$-i$
x_{27}	1	-1	1	$-i$
x_{28}	i	i	i	-1

$-\epsilon$

$$\begin{matrix} x_1 x_2 x_3 x_4 x_5 x_6 x_7 x_8 x_9 x_{10} x_{11} x_{12} x_{13} x_{14} x_{15} x_{16} x_{17} x_{18} x_{19} x_{20} x_{21} x_{22} x_{23} x_{24} x_{25} x_{26} x_{27} x_{28} \\ x_{18} x_{14} x_3 x_5 x_4 x_{13} x_{24} x_{20} x_9 x_{11} x_{10} x_{19} x_6 x_2 x_{15} x_{17} x_{16} x_1 x_{12} x_8 x_{21} x_{23} x_{22} x_7 - x_{28} - x_{27} - x_{26} - x_{25} \end{matrix}$$

γ

Denominator 5

	1	2	3	4	5	6	7	8	9	10	11	12	13	14	15	16	17	18	19	20	21	22	23	24	25	26	27	28
1	2	−2i	1	0	1	−2i	−1	0	0	i	1	0	0	0	−1	0	0	0	0	ī	1	i	−1	ī	0	−1	1	0
2	2i	−1	0	−1	2i	1	0	0	ī	−1	0	0	0	1	0	0	0	0	i	−1	ī	1	i	−2	i	0	ī	0
3	1	0	1	−2i	−1	0	0	i	1	0	0	0	−1	0	0	0	0	ī	1	i	−1	ī	2	−2i	1	−1	0	0
4	0	−1	2i	1	0	0	ī	−1	0	0	0	1	0	0	0	0	i	−1	ī	1	i	−2	2i	−1	0	i	ī	0
5	1	−2i	−1	0	0	i	1	0	0	0	−1	0	0	0	0	ī	1	i	−1	ī	2	−2i	1	0	1	0	−1	0
6	2i	1	0	0	ī	−1	0	0	0	1	0	0	0	0	i	−1	ī	1	i	−2	2i	−1	0	−1	ī	i	0	0
7	−1	0	0	i	1	0	0	0	−1	0	0	0	0	ī	1	i	−1	ī	2	−2i	1	0	1	−2i	0	1	−1	0
8	0	0	ī	−1	0	0	0	1	0	0	0	0	i	−1	ī	1	i	−2	2i	−1	0	−1	2i	1	ī	0	i	0
9	0	i	1	0	0	0	−1	0	0	0	0	ī	1	i	−1	ī	2	−2i	1	0	1	−2i	−1	0	−1	1	0	0
10	−i	−1	0	0	0	1	0	0	0	0	i	−1	ī	1	i	−2	2i	−1	0	−1	2i	1	0	0	0	ī	i	0
11	1	0	0	0	−1	0	0	0	0	ī	1	i	−1	ī	2	−2i	1	0	1	−2i	−1	0	0	i	−1	0	1	0
12	0	0	0	1	0	0	0	0	i	−1	ī	1	i	−2	2i	−1	0	−1	2i	1	0	0	ī	−1	i	ī	0	0
13	0	0	−1	0	0	0	0	ī	1	i	−1	ī	2	−2i	1	0	1	−2i	−1	0	0	i	1	0	0	−1	1	0
14	0	1	0	0	0	0	i	−1	ī	1	i	−2	2i	−1	0	−1	2i	1	0	0	ī	−1	0	0	i	0	ī	0
15	−1	0	0	0	0	ī	1	i	−1	ī	2	−2i	1	0	1	−2i	−1	0	0	i	1	0	0	0	1	−1	0	0
16	0	0	0	0	i	−1	ī	1	i	−2	2i	−1	0	−1	2i	1	0	0	ī	−1	0	0	0	1	0	i	ī	0
17	0	0	0	ī	1	i	−1	ī	2	−2i	1	0	1	−2i	−1	0	0	i	1	0	0	0	−1	0	1	0	−1	0
18	0	0	i	−1	ī	1	i	−2	2i	−1	0	−1	2i	1	0	0	ī	−1	0	0	0	1	0	0	ī	i	0	0
19	0	ī	1	i	−1	ī	2	−2i	1	0	1	−2i	−1	0	0	i	1	0	0	0	−1	0	0	0	0	1	−1	0
20	i	−1	ī	1	i	−2	2i	−1	0	−1	2i	1	0	0	ī	−1	0	0	0	1	0	0	0	0	ī	0	i	0
21	1	i	−1	ī	2	−2i	1	0	1	−2i	−1	0	0	i	1	0	0	0	−1	0	0	0	0	ī	−1	1	0	0
22	ī	1	i	−2	2i	−1	0	−1	2i	1	0	0	ī	−1	0	0	0	1	0	0	0	0	i	−1	0	ī	i	0
23	−1	ī	2	−2i	1	0	1	−2i	−1	0	0	i	1	0	0	0	−1	0	0	0	0	ī	1	i	−1	0	1	0
24	i	−2	2i	−1	0	−1	2i	1	0	0	ī	−1	0	0	0	1	0	0	0	0	i	−1	ī	1	i	ī	0	0
25	0	ī	1	0	1	i	0	i	−1	0	−1	ī	0	ī	1	0	1	i	0	i	−1	0	−1	ī	1	2	2	0
26	−1	0	−1	ī	0	ī	1	0	1	i	0	i	−1	0	−1	ī	0	ī	1	0	1	i	0	i	2	2	1	0
27	1	i	0	i	−1	0	−1	ī	0	ī	1	0	1	i	0	i	−1	0	−1	ī	0	ī	1	0	2	1	2	0
28	0	0	0	0	0	0	0	0	0	0	0	0	0	0	0	0	0	0	0	0	0	0	0	0	0	0	0	5

More recently Rudvalis, in work as yet unpublished, showed that there might be a rank 3 permutation group on 4060 points, of orbit lengths 1,1755,2304, the stabilizer being the twisted Ree group ${}^2F_4(2)$ which has a simple subgroup of index 2 (called the Tits group) of order 17,971,200. The group has been constructed by J. Conway and D. Wales [11].

This construction was not made in terms of permutations, but in terms of a matrix representation over the complex numbers. The covering group of the Rudvalis group over a center of order 2 has by its character table a representation of degree 28 and it was this which in fact they constructed. The program for these computations was written by C. Landauer. The writer found a new basis for the 28-dimensional representation of the Rudvalis group. This representation has the advantage, which the Conway-Wales representation did not have, that the matrices are all in the field $Q(i)$. In addition they are unitary. Generating matrices are shown in Table 5.9.

The matrix γ is given in full. The matrices α, β, and δ are given in abbreviated form. The matrix ϵ monomial. In this notation the elements $\alpha, -\beta^2, \gamma$ generate a group isomorphic to $L_2(25)$ which, extended by an automorphism of order 2, is the stabilizer of the Ree group F on the 2304 orbit. The matrices $\alpha, -\beta^2, \gamma, \delta\beta$, and ϵ generate the group F together with the center of order 2. Finally $\alpha, \beta, \gamma, \delta$, and ϵ (δ is redundant here) generate the covering group $2R$ of the simple Rudvalis group R. The 4060 points permuted by the Rudvalis group in its permutation form may be considered as 28-dimensional vectors over the ring $Z(i)$ of Gaussian integers in which the four vectors $\pm v, \pm iv$ correspond to the same point. In the notation of Table (5.9), the vector with 24 initial zeros $K_5 = (0, \ldots, 0, 3\text{-}i, 3\text{-}i, 3\text{-}i, 3\text{-}i)$ is one of the 4060 vectors and is stabilized by the group F above. In this notation the vectors are row vectors and if v is one of them and M a matrix of R, vM is the image of v under the action of M. In this representation the 4060 vectors are of norm 40 and given a vector v_0 whose stabilizer is F, the vectors of the 1755 orbit of F have inner product 0 with v_0 while those of the 2304 orbit have inner product $\pm 10, \pm 10i$. These vectors lie in a 28-dimensional lattice over $Z(i)$, and R is an automorphism group of this lattice.

In this matrix representation the known group F moved vectors of norm 40 in orbits of lengths 1, 2304, 1755. Under the action of the subgroup L these split into orbits of lengths

$$1 + (1 + 78 + 300 + 300 + 325 + 325 + 975) + (975 + 390 + 390) \quad (5.10)$$

Representatives of these orbits were found. Under the action of β, since β normalizes L it is only necessary to check that β takes one vector from each of these orbits into a vector of some orbit. This shows that $2R = \langle\alpha, \beta, \gamma, \delta, \epsilon\rangle$ is transitive on the 4060 points and proves the existence of the group. The associated graph joins each point to the 1755 points of the 1755 orbit in its stabilizer. Conway and Wales were able to show that R was the full automorphism group of the graph and in particular R has no outer automorphisms.

The Conway group [10] was given initially as a group of 24-dimensional real orthogonal matrices, the automorphism group (excluding translations) of the Leech lattice. The minimum length of a vector in this lattice is $\sqrt{32}$ and there are 196,560 vectors of this length in the lattice. The main initial problem which Conway faced was to find matrices which would generate a group transitive of these vectors.

In 1956 Z. Janko [20] proved the following theorem:

Theorem [Janko]. *Let* G *be a finite group with the following properties:*
(*a*) *Sylow* 2*-subgroups of* G *are Abelian,*
(*b*) G *has no subgroup of index* 2,
(*c*) G *contains an involution* t *such that*

$$C_G(t) = \langle t \rangle \times F, \text{ where } F \cong A_5.$$

Then G *is a new simple group* J *of order* 175,560.

The group J is generated by the two following 7-dimensional matrices over $GF(11)$.

$$A = \begin{bmatrix} 0 & 1 & 0 & 0 & 0 & 0 & 0 \\ 0 & 0 & 1 & 0 & 0 & 0 & 0 \\ 0 & 0 & 0 & 1 & 0 & 0 & 0 \\ 0 & 0 & 0 & 0 & 1 & 0 & 0 \\ 0 & 0 & 0 & 0 & 0 & 1 & 0 \\ 0 & 0 & 0 & 0 & 0 & 0 & 1 \\ 1 & 0 & 0 & 0 & 0 & 0 & 0 \end{bmatrix} \quad B = \begin{bmatrix} -3 & 2 & -1 & -1 & -3 & -1 & -3 \\ -2 & 1 & 1 & 3 & 1 & 3 & 3 \\ -1 & -1 & -3 & -1 & -3 & -3 & 2 \\ -1 & -3 & -1 & -3 & -3 & 2 & -1 \\ -3 & -1 & -3 & -3 & 2 & -1 & -1 \\ 1 & 3 & 3 & -2 & 1 & 1 & 3 \\ 3 & 3 & -2 & 1 & 1 & 3 & 1 \end{bmatrix} \tag{5.11}$$

Using character theorems due to M. Suzuki, Janko was first able to show that his hypotheses forced the group to be of order 175,560 and also simple. Then he found its full character table. From modular theory he was able to show that group J, if it existed, would have to have a 7-dimensional representation over $GF(11)$.

A Sylow 2-subgroup S of J is elementary Abelian of order 8 and its normalizer H_1 is of order 168. H_1 was easily constructed. Here $S = \langle t_1, t_2, t_3 \rangle$ where t_1, t_2, t_3 are commuting involutions. Further elements x, y satisfy:

$$\begin{gathered} x^3 = y^7 = 1, x^{-1}yx = y^2 \\ y^{-1}t_1y = t_2, y^{-1}t_2y = t_3, y^{-1}t_3y = t_1t_3 \\ x^{-1}t_1x = t_1, x^{-1}t_2x = t_3, x^{-1}t_3x = t_1t_2t_3 \end{gathered} \tag{5.12}$$

Then $H_1 = \langle x, y, t_1 \rangle$. He can show that a further involution t_4 exists such that

$t_4 x = x t_4$ and $t_4 y\, t_4 = y^{-1}$ and $(t_1 t_4)^5 = 1$. Only one matrix exists with properties required of t_4. Then there are 11 double cosets $H_1 C_i H_1$ and calculations show that these 11 cosets together form the group J.

John Lindsey [21] has shown that the Hall-Janko group extended by a center of order 2 has an irreducible complex representation of degree 6. This representation can be realized over the field $Q(\sqrt{5}, \sqrt{-7})$. It is however more convenient to consider these over finite fields. Here HJ = $\langle a, u, t\rangle$ and representations over $GF(5)$ and $GF(9)$ are given

Over $GF(5)$

$$a = \begin{bmatrix} 1 & 1 & 1 & -2 & 1 & 1 \\ -2 & -1 & -2 & 1 & 2 & -2 \\ 0 & 0 & -2 & 1 & 0 & 2 \\ -2 & 0 & -2 & 2 & -2 & -2 \\ 1 & -1 & -2 & 2 & -1 & -1 \\ 2 & 1 & 1 & 0 & -2 & 0 \end{bmatrix} \quad u = \begin{bmatrix} -2 & 0 & 0 & 1 & 0 & 0 \\ 0 & 0 & 0 & 0 & 1 & 0 \\ 0 & 0 & -1 & 0 & 0 & -1 \\ 2 & 0 & 0 & 1 & 0 & 0 \\ 0 & -1 & 0 & 0 & -1 & 0 \\ 0 & 0 & 1 & 0 & 0 & 0 \end{bmatrix}$$

$$t = \begin{bmatrix} 0 & 0 & 0 & 1 & 0 & 0 \\ 0 & -2 & -2 & 0 & 1 & -2 \\ 0 & 0 & 2 & 0 & -2 & -1 \\ -1 & 0 & 0 & 0 & 0 & 0 \\ 0 & -1 & 2 & 0 & 2 & 0 \\ 0 & 2 & 1 & 0 & 2 & -2 \end{bmatrix} \tag{5.13}$$

Over $GF(9)$

$$a = \begin{bmatrix} i & 1-i & -i & 0 & 0 & 0 \\ 1-i & -i & -1 & 0 & 0 & 0 \\ -i & -1 & 1+i & 0 & 0 & 0 \\ 0 & 0 & 0 & -i & 1+i & i \\ 0 & 0 & 0 & 1+i & i & -1 \\ 0 & 0 & 0 & i & -1 & 1-i \end{bmatrix} \quad u = \begin{bmatrix} 1+i & -i & i & 0 & 0 & 0 \\ -i & 1+i & -i & 0 & 0 & 0 \\ i & -i & 1+i & 0 & 0 & 0 \\ -i & 0 & 0 & 1-i & i & -i \\ 0 & -i & 0 & i & 1+i & i \\ 0 & 0 & -i & -i & i & 1+i \end{bmatrix}$$

$$t = \begin{bmatrix} 0 & 0 & 0 & i & 0 & 0 \\ 0 & 0 & 0 & 0 & i & 0 \\ 0 & 0 & 0 & 0 & 0 & i \\ i & 0 & 0 & 0 & 0 & 0 \\ 0 & i & 0 & 0 & 0 & 0 \\ 0 & 0 & i & 0 & 0 & 0 \end{bmatrix}$$

6. Investigations on Burnside groups

A Burnside group $B(n, r)$ is a group generated by r elements $a_1, a_2, \ldots, a_r$ with the defining relations $x^n = 1, x \in B(n, r)$. It is known that $B(n, r)$ is finite for $n = 1, 2, 3, 6$ and any finite r. It is also known, from the work of Adyan and Novikov that $B(n, 2)$ is infinite if n is odd and sufficiently large. No attempt will be made here to review the extensive literature on Burnside groups. Only computer attacks on these groups will be mentioned.

At first glance coset enumeration might seem appropriate, but this seems to have very limited application. For example, for exponent 4 we may take $B(4, 2) = \langle a, b \rangle$ and with relations

$$\begin{gathered} a^4 = b^4 = (ab)^4 = (a^{-1}b)^4 = (a^2b)^4 = (a^2b^2)^4 = 1 \\ (abab^{-1})^4 = 1, (a^{-1}b^{-1}ab)^4 = 1 \end{gathered} \tag{6.1}$$

Enumerating G on cosets of $H = \langle a, b^2 \rangle$ we find 64 cosets and also find H to be of order 64 so that we may have shown that the order of $G = B(4, 2)$ is at most 2^{12} providing that H does not contain a proper normal subgroup of G. Indeed 2^{12} is the order of $B(4, 2)$ and H is of order 64, as can be shown with further work. But the groups, even when known to be finite, are of such huge orders that they go beyond the range of the computer storage available except in the simplest cases.

A program by A. L. Tritter [25] was designed to decide whether or not, in the Burnside group G of exponent 4 with 8 generators $\langle g_0, g_1, g_2, g_3, g_4, g_5, g_6, g_7 \rangle$, the third derived group would belong to G_9 the 9th term in the lower central series. The calculations were carried out in the corresponding Lie ring. The free Lie ring has a basis of $7! = 5040$ elements over $GF(2)$. Relations on the ring are a consequence of the commutator collection process on products $(x_1, \ldots, x_8)^4 = 1$ where $x_1, \ldots, x_8$ are a permutation of $g_0, \ldots, g_7$. It is not known to the writer whether this program was carried to completion or not.

Computer computations on groups of exponent 4 have been carried out by Bachmuth, Mochizuki, and Weston [3]. They investigate groups H of exponent 4 with a possibly countable number of generators $g_i, i = 1, \ldots$ such that

(*a*) $g_i^2 = 1$,
(*b*) the normal closure of g_i in H is Abelian,
(*c*) $(g_i, h, h, h) = 1$ for all h in H.

They prove that H''', the third derived group of H, is not the identity. The properties (*a*), (*b*), (*c*) assure that the group is of exponent 4. Gupta and Weston [13] have reduced the question of the derived length of groups $B(4, r)$ to that of groups of type H. Bachmuth *et al.* reduce this to study of a group over a free associative ring R with identity over $GF(2)$ generated by non-commuting indeterminates $x_1, x_2, \ldots$.

In R we define a group H^*, a homomorphic image of H above with $g_i \to 1 + x_i$ and conditions (a), (b), (c). Define $T_3(x_1, \ldots, x_n)$ as the homogeneous component of degree one in each indeterminate of $[(1 + x_1)(1 + x_2) \ldots (1 + x_n) - 1]^3$. Write $\Delta(X, Y) = XY + YX$. Then with $\Delta_1(x_1, x_2) = x_1 x_2 + x_2 x_1$ we define in turn

$$\Delta_2(x_1, x_2, x_3, x_4) = \Delta(\Delta(x_1, x_2), \Delta(x_3, x_4))$$

$$\Delta_3(x_1 \ldots x_8) = \Delta(\Delta_2(x_1, x_2, x_3, x_4), \Delta_2(x_5, x_6, x_7, x_8))$$

It will follow that $H''' \neq 1$ if it can be shown that $\Delta_3(x_1, \ldots, x_8)$ is not in the ideal generated by the T_3's. This they put onto the computer and showed that $\Delta(x_1, x_5, x_2, x_7, x_3, x_6, x_4, x_8)$ is not in the ideal even if the further homomorphism φ is applied

$$\varphi: x_1 \to x_1, \quad x_2 \to x_1, \quad x_3 \to x_2, \quad x_4 \to x_2, \quad x_5 \to x_3,$$
$$x_6 \to x_3, \quad x_7 \to x_4, \quad x_8 \to x_4$$

Using programs written by Donald Knuth the writer began investigations on $B(4, 3) = G$ by finding generators for G' and found the relations on these generators in terms of the generators for G. This has never been completed, but might be resumed some day. Further programs were devised to express arbitrary commutators of weight n modulo G_{n+1} in terms of basic commutators.

An interesting problem is the group $B(5, 2)$ which if it is finite has order 5^{33} or 5^{34}. The technical problems in studying $B(5, 2)$ would be considerable, but it would be a desirable first step in the investigation of groups of exponent 5. The writer conjectures that $B(5, r)$ is always finite and even of order less than that of $B(6, r)$ which is known to be of order $2^A 3^B$, $A = 1 + (r-1) . (3^s)$,

$$s = r + \binom{r}{2} + \binom{r}{3}, B = t + \binom{t}{2} + \binom{t}{3}, t = 1 + (r-1) 2^r.$$

References

1. M. Aschbacher and M. Hall, Jr. Groups generated by a class of elements of order 3. *J. Algebra* **24**(1973), 591–612.
2. J. Assion. "Eine Kennzeichnung der alternienden Gruppen". Diplomarkeit, Frankfurt, 1970.

3. S. Bachmuth, H. Y. Mochizuki, and K. Weston. A group of exponent 4 with derived length at least 4. *Proc. Amer. Math. Soc.* **39**(1973), 228–234.
4. R. Brauer. On groups whose order contains a prime number to the first power, I. II. *Amer. J. Math.* **64**(1942), 401–440.
5. R. Brauer and W. F. Reynolds. On a problem of E. Artin. *Ann. of Math.* **68**(1958), 713–720.
6. R. Brauer and K. A. Fowler. On groups of even order. *Ann. of Math.* **62**(1955), 565–583.
7. R. Brauer and Chi-Han Sah. "Theory of Finite Groups". Benjamin, New York and Amsterdam, 1969.
8. J. S. Brodkey. A note on finite groups with an Abelian Sylow group. *Proc. Amer. Math. Soc.* **14**(1963), 132–133.
9. W. Burnside. "The Theory of Groups". 2nd Edition, Cambridge University Press, 1911.
10. J. H. Conway. A group of order 8,315,553,613,086,720,000. *Bull. London Math. Soc.* **1**(1969), 79–88.
11. J. Conway and D. Wales. Construction of the Rudvalis group of order 145,926,144,000. *J. Algebra* **27**(1973), 538–548.
12. H. S. M. Coxeter and J. A. Todd. Abstract definitions for the symmetry groups of the regular polytopes in terms of two generators. Part I: The complete groups. *Proc. Cambridge Philos. Soc.* **32**(1936), 194–200.
13. N. D. Gupta and K. Weston. On groups of exponent four. *J. Algebra* **17**(1971), 59–66.
14. Marshall Hall, Jr. A search for simple groups of order less than one million. *In* "Computational Problems in Abstract Algebra", Pergamon Press, Oxford and New York, 1969, pp. 137–168.
15. Marshall Hall, Jr. Construction of finite simple groups. *In* "Computers in Algebra and Number Theory", SIAM-AMS Proceedings, Vol. IV (1971).
16. Marshall Hall, Jr. On the number of Sylow subgroups in a finite group. *J. Algebra* **7**(1967), 363–371.
17. D. G. Higman. Finite permutation groups of rank 3. *Math. Zeitschr.* **86**(1964), 145–156.
18. D. G. Higman and C. Sims. A simple group of order 44,352,000. *Math. Zeitschr.* **105**(1968), 110–113.
19. Graham Higman and John McKay. On Janko's simple group of order 50,232,960. *Bull. London Math. Soc.* **1**(1969), 89–94.
20. Z. Janko. A new finite simple group with Abelian Sylow 2-subgroups and its characterization. *J. Algebra* **3**(1966), 147–186.
21. J. H. Lindsey II. On a projective representation of the Hall-Janko group. *Bull. Amer. Math. Soc.* **74**(1968), 1094.
22. John McKay and David Wales. The multipliers of the simple groups of order 604,800 and 50,232,960. *J. Algebra* **17**(1971), 262–272.
23. R. G. Stanton. The Mathieu groups. *Can. J. Math.* **3**(1951), 164–174.

24. B. Stellmacher. Einfache Gruppen, die von einer Konjugiertenklasse von Elementen der Ordnung drei erzeugt werden. *J. Algebra* **30**(1974), 320–354.
25. A. L. Tritter. A module-theoretic computation related to the Burnside problem. "Computational Problems in Abstract Algebra", Pergamon Press, Oxford and New York, 1970.

2. Computer Proof of Relations in Groups

JOHN LEECH

Department of Computing Science, University of Stirling, Stirling, Scotland

1. Introduction

In 1936, Todd and Coxeter gave a method ([15], also described in [6], Chapter 2) for establishing the index, when finite, of a subgroup of a finitely presented group. Their principal application was to examples where the subgroup is of known finite order, thus exhibiting the orders of the whole groups. Their method consists of enumerating systematically the cosets of the subgroup generated by elements given as words in the generators of the group. The method was described as "purely mechanical", and the advent of electronic computers has led a number of people to program the method for automatic execution. I have given accounts of this work [8, 11], and Cannon *et al.* have given a detailed comparative analysis of computer methods [2].

A proof by coset enumeration (whether by hand or by computer) that a group is finite clearly implies proof of many relations in the group, for example those specifying periods of elements of the group. But it does not readily yield explicit proofs of these relations, particularly when the work is done out of sight on a computer and only the results are presented. This is conspicuous in cases where the group is trivial; computer working commonly destroys its own traces, and one is left with no supporting evidence that the group is trivial to supplement the bare statement that "the computer says so". Since almost any sporadic computer malfunction, and many a subtle program error, will precipitate a collapse of the working to a single coset, one's confidence in such a conclusion may well be less than complete. Those performing hand work may be similarly deceived; Coxeter ([3], footnote p. 143) reported independent coset enumerations, by Sinkov and himself, purporting to show that the group ((5, 5, 5; 3)) was trivial, yet later work [14, 5] showed that this conclusion was wrong (see Section 12).

The object of the present work has been to derive extensions to the Todd-Coxeter method which will allow the user to deduce formal proofs of relations whose proofs are implicit in the working of the coset enumeration, and as far as possible to have these proofs derived by the computer itself. My first essay in this direction [7] was applied by hand to cases where the coset enumeration had not involved definition and subsequent elimination of redundant cosets. Later [9] I programmed part of the working, but it was still necessary to prepare by hand a suitably annotated coset multiplication table, which was then put into the computer along with other data pertaining to the relations to be proved. It was still necessary for the enumeration to have been performed without redundant cosets having been defined, which made it unsuitable for full computer implementation. A later adaptation of the method [11] overcame this restriction, and in the present account I describe the method and its computer implementation and give some examples of work that has been done with it. The account concludes with a description of a further adaptation to give Reidemeister-Schreier relations for the subgroup, in a way that does not involve the introduction of new generators.

2. Coset enumeration

This is not the occasion for a detailed description of coset enumeration (see references cited in Section 1). In hand work, each relator in the presentation of a group is spelt out at the head of a table, whose columns fall below the spaces between the letters of the relator. Each row of each table is filled with coset numbers, beginning and ending with the same coset, the cosets in adjacent columns being related by the generator at the head of the space between the columns. It is convenient also to keep a multiplication table showing the effect on each coset of each generator, and also of the inverse of each generator which is not involutory, as these are determined. Initially entries are made so as to define the subgroup which is chosen as coset 1. As the work progresses, more cosets are defined and entered in the tables, and whenever a line of a relator table is completed, further entries are made in the multiplication and other tables. The enumeration is complete when the tables are full, leaving no space for the definition of further cosets, and exhibiting every coset in every significantly different position in each relator table.

The cosets, and the effects on them of multiplication by generators of the group, may be represented by a graph. The nodes of the graph are the cosets, and the edges join pairs of cosets related by generators. Edges are normally directed, though for involutory generators it is convenient to use a single undirected edge in place of a pair of directed edges joining a pair of nodes in both directions. It is convenient (to all except printers) to use colours for distinguishing different generators. This graph is called a *Schreier diagram.* If the subgroup is trivial, the graph is the *Cayley diagram* for the group, while if the subgroup is normal, the graph is the Cayley diagram for the factor group obtained by factoring out the subgroup. Of course it is feasible to draw the graph only if the number of cosets is not too large, perhaps a few hundreds at the most.

3. The basic method

Given an element of a group, presented as a word in the generators of the group, and an enumeration of the cosets of a subgroup of the group, we can readily determine whether the element is an element of the subgroup. All we have to do is to begin with coset 1 and multiply by the successive letters of the word; the element is an element of the subgroup if and only if the final result is coset 1. It may be convenient to perform this multiplication by tracing the appropriate path on the Schreier diagram for the cosets. However, although this will exhibit which elements of the group are elements of the subgroup, it does not enable us to exhibit them as words in the subgroup generators. The present method was devised to allow derivation of the appropriate words in the subgroup generators.

The Schreier diagram is not a tree, as it has circuits in it. These are introduced in two ways. First, there are circuits beginning and ending at coset 1 which correspond to the elements generating the subgroup. Then there are circuits corresponding to complete lines of working in the relator tables. For each of these latte whose completion was used to deduce an entry in the multiplication table, we may characterise the entry as a deduced entry, and the corresponding edge as a deduced edge. If we were to remove all these deduced edges from the graph, it would become almost a tree, having in it only circuits corresponding to subgroup generators. (It would become a tree when the enumeration is of cosets of the trivial subgroup, i.e. of elements of the group.) In particular, there would be a path, unique except for ambiguity due to these remaining circuits, from each coset back to coset 1, corresponding to its definition. We shall exhibit a method for examining any circuit, beginning and ending at coset 1, and eliminating from it any deduced edges. The result is a circuit, beginning and ending at coset 1, which is made up of a sequence of circuits corresponding to subgroup generators. This represents the expression of an element of the subgroup, given as a word in the generators of the group, as a word in the generators of the subgroup. The method is illustrated in the following example.

4. A worked example

In this section I enumerate the six cosets of the subgroup $\{A\}$ of the octahedral group defined by the relations

$$A^4 = B^3 = (AB)^2 = I.$$

Computer implementation is described in the next section; however, I anticipate the practice of later sections by using barred letters $\bar{A}$, $\bar{B}$ for inverses of non-involutory generators. (Because of the restricted character sets available, I use other letters for inverses in computer input and output; as these are less perspicuous and no more compact than barred letters, they are not used in this account.)

First, the subgroup is specified by making the entries $1A = 1$, $1\bar{A} = 1$ in the multiplication table. Next we define cosets $2 = 1B$ and $3 = 1\bar{B}$. On inserting these entries into the relator tables, we find that two lines of working are closed, enabling us to deduce $2B = 3$ from $B^3 = I$ and $2A = 3$ from $(AB)^2 = I$. Corresponding to each of these we record deduction words

$$2B3 = \bar{B}\bar{B} \tag{1}$$

and

$$2A3 = \bar{B}\bar{A}\bar{B} \tag{2}$$

the words on the right corresponding to paths in the Schreier diagram joining the same cosets as the deduced edges but not themselves containing deduced edges. Next we define $2\bar{A} = 4$ and $3A = 5$. We can now deduce $5A = 4$ from $A^4 = I$ and $5B = 4$ from $(AB)^2 = I$. Now, however, the deduction words are less straightforward. $5A = 4$ gives the word $\bar{A}\bar{A}\bar{A}$; letters are related by $5\bar{A} = 3$, $3\bar{A} = 2$, $2\bar{A} = 4$, of which $3\bar{A} = 2$ is deduced and is to be replaced in turn by BAB ((2) above), and we obtain

$$5A4 = \bar{A}BABA\bar{A} \tag{3}$$

Similarly $5B = 4$ gives the word $\bar{A}\bar{B}\bar{A}$; the letter $\bar{B}$ relates $3\bar{B} = 2$ which is to be replaced by BB ((1) above), so we obtain

$$5B4 = \bar{A}BB\bar{A} \tag{4}$$

Lastly we define $4B = 6$, and deduce $6B = 5$ from $B^3 = I$ and then $6A = 6$ from $(AB)^2 = I$. The corresponding deduction words, obtained as above, are

$$6B5 = \bar{B}A\bar{B}\bar{B}A \tag{5}$$

$$6A6 = \bar{B}A\bar{B}\bar{A}B\bar{A}B \tag{6}$$

The work is now complete, and we have the multiplication and relator tables given below. The small figures in the relator tables denote the deductions made by the closures of these lines, numbered as above in the sequence in which they were made. The same figures appear as subscripts to the corresponding entries in the multiplication table.

	A	$\bar{A}$	B	$\bar{B}$
1	1	1	2	3
2	3_2	4	3_1	1
3	5	2_2	1	2_1
4	2	5_3	6	5_4
5	4_3	3	4_4	6_5
6	6_6	6_6	5_5	4

	A	A	A	A
1	1	1	1	1
2	3	5 $_3$	4	2
6	6	6	6	6

	A	B	A	B
1	1	2 $_2$	3	1
3	5 $_4$	4	2	3
5	4	6 $_6$	6	5

	B	B	B
1	2 $_1$	3	1
4	6 $_5$	5	4

The Schreier diagram for this enumeration is given below.

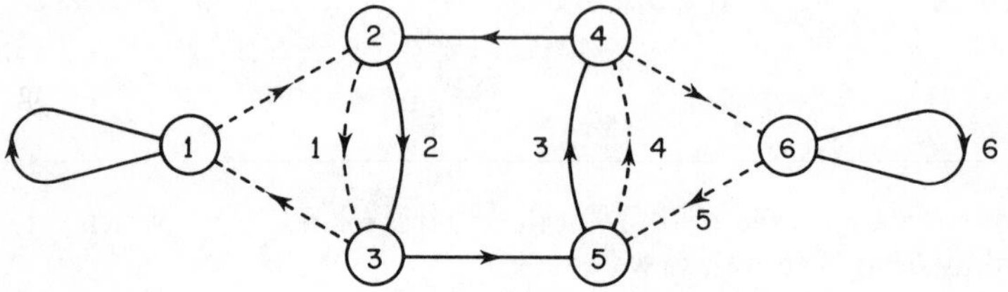

In this diagram the ringed numbers are the cosets, the solid edges join cosets related by A and the broken edges those related by B. The numbered edges are deduced, and are numbered to correspond with the deductions above. It is easily checked that for each deduced edge the deduction word obtained above joins the same two nodes by a path using no deduced edge, while the deduction procedure ensures that the deduction word is the same element as the letter it replaces.

I now illustrate the use of this in proving relations. For example, suppose we wish to ascertain the period of the element $C = \bar{A}B$. By reference to the coset multiplication table, or the Schreier diagram, we find that $1C = 2$, $2C = 6$, $6C = 5$, $5C = 1$, so we know that $1C^4 = 1$. Thus C^4 is an element of the subgroup, i.e. a power of A; as yet we have not found what power. So we write C^4 in full and fill in coset numbers underneath:

$$\begin{array}{lccccccccc} C^4 = & \bar{A} & B & \bar{A} & B & \bar{A} & B & \bar{A} & B & \\ & 1 & 1 & 2 & 4 & 6_6 & 6_5 & 5 & 3 & 1 \end{array}$$

The small figures denote that the relations $6\bar{A} = 6$, $6B = 5$ were deduced, deductions 6 and 5 respectively. Now we substitute the deduction words, and obtain

$$C^4 = \bar{A}B\bar{A}B.\bar{B}A\bar{B}AB\bar{A}B.\bar{B}A\bar{B}\bar{B}A.\bar{A}B$$

$$= I$$

establishing the period as required. The reader may verify similarly that the period of the commutator $D = \bar{A}\bar{B}AB$ is 3; in the first instance one obtains $D^3 = A^{-4}$, and reference to the defining relation $A^4 = I$ is needed to complete the proof. This illustrates that when the element is expressed as a word in the subgroup generators, it is not necessarily derived in its simplest form.

5. Computer implementation

It is not convenient on a computer to keep tables for each relator; instead, each entry placed in the multiplication table, by definition or deduction, is tried in each significantly different position in each relator, as much as possible of each such line of working being calculated. Any further entry deduced by closure of a line of working is stored in a list of unworked table entries to be tried in its turn. The main working table is the coset multiplication table, which has provision for storing the deduction number of each deduced entry. The corresponding deduction word is stored in a separate list; since words are of unequal length, it is convenient to keep a key list which holds the addresses of the beginning and end of each word. Two mutually inverse deduced entries in the multiplication table are distinguished by sign; thus, in the example of Section 4, the first deduction was $2B = 3$, with word $\bar{B}\bar{B}$, which would be numbered 1, while $3\bar{B} = 2$ would be labelled -1 to indicate that the deduction word is the inverse of word 1 and is not stored separately.

The deduction words are close packed with several letters per computer word, and are unpacked as required. A new deduction word is formed by spelling out the rest of the relator from which the multiplication table entry was deduced, examining the coset numbers to find which letters correspond to deduced entries, unpacking and substituting the deduction words for these letters, closing up the resulting word by cancelling any letter and its inverse adjacent to it, and close packing the resulting word for storage.

The computer output includes the following items, some optional. First there is a copy of the data, comprising the relators and subgroup definition. Next follows, optionally, a record of the working as it is done. The example of Section 4 would appear as follows.

DEFINE	$1\ B\ 2$					
DEFINE	$1\ \bar{B}\ 3$					
1	$2\ B\ 3$	BY	3	FROM	BBB	$\bar{B}\bar{B}$
2	$2\ A\ 3$	BY	3	FROM	$BABA$	$\bar{B}\bar{A}\bar{B}$
DEFINE	$2\ \bar{A}\ 4$					
DEFINE	$3\ A\ 5$					
3	$5\ A\ 4$	BY	3	FROM	$AAAA$	$\bar{A}BAB\bar{A}$
4	$5\ B\ 4$	BY	3	FROM	$ABAB$	$\bar{A}BB\bar{A}$
DEFINE	$4\ B\ 6$					
5	$6\ B\ 5$	BY	4	FROM	BBB	$\bar{B}A\bar{B}\bar{B}A$
6	$6\ A\ 6$	BY	4	FROM	$BABA$	$\bar{B}A\bar{B}\bar{A}B\bar{A}B$

In each line listing a deduced entry, the line of working of the relator table, in the cyclic permutation shown, beginning with the coset following BY, is that used to make the deduction. The first letter in the relator as shown relates cosets according to the multiplication table entry being currently examined; in this example two deductions were made from each of the entries $3B1$, $3A5$, $4B6$.

Further output includes the final number of cosets, and optionally the final multiplication table with numbers for the deduced entries and a list of the deduction words. Following these, we can read words in the group generators, which will be checked for membership of the subgroup and the corresponding elements printed out expressed as elements in the subgroup.

6. Coincidences

So far, our performance of Hamlet has avoided any reference to the Prince of Denmark. It is unnecessary, and often impossible, to find a sequence of definitions of cosets which will allow the whole of an enumeration to be performed without the definition of apparently new cosets which are later found to coincide with cosets already defined. A comprehensive computer program must be able to deal with these when they occur; only then will it be widely useful. It is then unnecessary to go to great lengths to seek sequences of definitions of cosets which avoid these coincidences; it is sufficient to adopt a uniform procedure, such as that adopted in the present programs which is to make each new definition so as always to fill the earliest blank space in the coset multiplication table.

Coincidences are discovered when, in the course of constructing a line of a relator table, it is found that the working overlaps, indicating that two different numbers should appear in the same place, so that these numbers designate the same coset. The coset with the greater number is then deleted, and references to it are replaced by references to the smaller number. This may lead to further coincidences, either directly because the products by the same generator of the two cosets are already defined and have different numbers, or indirectly because further lines of relator tables are found to overlap. In some cases the working leads to a total collapse, all cosets being found to coincide. Here one's main interest is in the working, leading to a formal proof that the subgroup generators generate the whole group, which may imply that the group is trivial. In other cases it is found that there is a partial collapse, a substantial number of cosets being eliminated to leave a consistent set of cosets. This is sometimes due to the existence of a relation which, although not necessary to the presentation of the group, is unusually difficult to derive from the defining relations, and whose adjunction to them simplifies the enumeration considerably. Or again there may be only a few cosets eliminated. This may be due to the sequence of definitions being inefficient, the coincidences being avoidable with care and personal attention.

7. Computer handling of coincidences

The correct handling of coincidences forms the most complex part of programs for coset enumeration, and this is no less true when the process is being adapted for proving relations. First I describe the procedure for handling coincidences without reference to proving relations.

Suppose it has been found that cosets numbered a, b are the same, where $b > a$. We have to replace references to b in the multiplication table by references to a. We examine each entry in line b of the multiplication table. If it is blank, no action is taken and we move on to the next entry. If we find an entry $bR = b$ in the column for generator R, then we replace this by a. But if we find an entry $bR = c \neq b$, then we know that line c contains an entry $c\bar{R} = b$. In the first instance we delete this rather than replace it by a, to avoid having two occurrences of a in the same column. Then, whether $bR = b$ or not, we examine aR. If this was not defined, we copy bR there, and place this entry in the list of unworked table entries. But if it was defined, then if $aR = b$ we replace it by a, and in any case we set up a new coincidence between bR and aR, and place it in a list of coincidences to be dealt with. In either case, if $(aR)\bar{R}$ is undefined, we set it equal to a.

Any further coincidence discovered in this way, by bR and aR being defined and unequal, is placed in a list of coincidences to be dealt with. The following way of keeping this list is to ensure that it never contains redundant information. Each entry is a pair of numbers, the greater preceding the lesser, stored in decreasing order of these greater numbers. When a new coincidence is to be placed in the list, a search is made to find whether its greater number is the greater number of any pair in the list. If not, the pair is placed in the list at the appropriate place. But if the greater number occurs in the list, then the lesser numbers of the old and new pairs are paired and replace the original pair in the search of the list. Eventually either a pair is formed whose greater number does not equal that of any pair in the list, which we then place in the list, or a pair is formed of two equal numbers, indicating that this coincidence gave no new information, and no entry is placed in the list.

When a coincidence has been fully dealt with as above, the next coincidence, at the head of the list to be dealt with, is treated similarly, and we continue in this way until this list is empty. We then return to normal working, beginning with the list of unworked entries in the multiplication table, continuing until this is empty or another coincidence occurs. If in this list any entry relates cosets, one (or both) of which has been deleted by coincidence, it is ignored, as the corresponding working will be done from other entries not thus ignored. When this list is exhausted, and before defining any new cosets, the multiplication table is closed up by transferring and renumbering cosets so as not to leave gaps from which cosets have been deleted. Then if there are still any undefined cosets we return to defining new cosets, otherwise the working is complete and the results are ready for output.

8. Coincidences and word construction

So far I have not referred to the construction of deduction words for multiplication table entries which have been made or altered when dealing with coincidences. This is done in the following way.

When an original coincidence is found (i.e. one which is not obtained as a consequence of another coincidence), there is a place in a relator table which can be filled by two different coset numbers. Thus there is a cyclic permutation of the relator such that we can begin with one coset, work right through the relator, and end with a different coset. From this line of working we form a coincidence word, in much the same way as a deduction word, by replacing any letters in the relator corresponding to deduced multiplication table entries by the deduction words. This is numbered and stored similarly to deduction words. If the working is being printed out, an original coincidence is printed similarly to a deduction.

For a derived coincidence, when bR and aR are both defined, the coincidence word is formed from the former coincidence word bIa by adjoining $\bar{R}$ and R to form $\bar{R}bIaR$, noting that if bR or aR is a deduced entry the corresponding deduction word is to be substituted for $\bar{R}$ or R. Also if aR or bR is b and is to be replaced by a, the coincidence word bIa is to be adjoined. Thus if $bR = b$ and $aR = c$, we get $aIc = a\bar{I}b\bar{R}bIaRc$ for the coincidence word, with appropriate substitutions on the right hand side.

If a derived coincidence is placed directly in the coincidence list, the coincidence word is stored as formed above. However, if in trying to place a coincidence between b and a in the list we find a coincidence between b and c in the list already, we form the new pair a, c with coincidence word $aIc = a\bar{I}bIc$ and continue the search. This may happen repeatedly.

When bR is defined but aR is not, we assign to aR the value of bR. If $bR = c \neq b$, the deduction word for aRc is $a\bar{I}bRc$. If $bR = b$, the deduction word for aRa is $a\bar{I}bRbIa$, the coincidence word bIa being adjoined twice.

Of course, when these deduction and coincidence words are being adjoined in this way, whenever any letter is adjacent to its inverse these are cancelled. Thus before it is stored any such word is reduced as much as possible.

When a coincidence and all its consequences have been dealt with, the coset multiplication table is closed up to eliminate space left by deleted cosets, as stated in Section 7. Additionally all coincidence words and all deduction words corresponding to deleted multiplication table entries are deleted at this stage as they are not required for further working. If the working is being printed out, they will be available for inspection, and the entire working can be reconstructed if required from this printed record.

9. Example to illustrate working

This is a group given by Conway, with five generators and five relations, each relation equating a generator to the product of the two generators preceding it in cyclic order. Thus the relators are $AB\bar{C}$, $BC\bar{D}$, $CD\bar{E}$, $DE\bar{A}$, $EA\bar{B}$. We show that this group is cyclic by establishing that $\{A\}$ has only one coset. Having defined coset 1 by $1A = 1$, we begin the following working:

DEFINE	$1\ B\ 2$					
1	$1\ C\ 2$	BY	1	FROM	$B\bar{C}A$	AB
DEFINE	$1\ \bar{B}\ 3$					
2	$3\ D\ 2$	BY	3	FROM	$BC\bar{D}$	BAB
3	$3\ E\ 1$	BY	1	FROM	$\bar{B}EA$	$B\bar{A}$

4	1 D 3	BY	3	FROM	$E\bar{A}D$	$AA\bar{B}$
5	2 C 3	BY	3	FROM	$\bar{D}BC$	$\bar{B}AA\bar{B}$
6	2 E 2	BY	2	FROM	$CD\bar{E}$	$\bar{B}AAAB$
7	3 A 2	BY	2	FROM	$E\bar{A}D$	$BAAAAB$
8	1 E 3	BY	3	FROM	$A\bar{B}E$	$\bar{A}\bar{A}\bar{A}\bar{A}\bar{B}$
9	2 = 1	BY	3	FROM	$\bar{E}CD$	$\bar{B}\bar{A}\bar{A}\bar{A}\bar{A}\bar{A}\bar{A}\bar{A}$
10	3 = 1					$B\bar{A}\bar{A}\bar{A}\bar{A}$
11	1 C 1					$\bar{A}\bar{A}\bar{A}\bar{A}\bar{A}\bar{A}$
12	3 D 1					$B\bar{A}\bar{A}\bar{A}\bar{A}\bar{A}\bar{A}$
13	1 E 1					AAA
14	1 B 1					$AAAA$
15	1 D 1					$\bar{A}\bar{A}$

It is instructive to examine the multiplication table as it is after deduction 8, before the collapse; it is then as follows:

	A	$\bar{A}$	B	$\bar{B}$	C	$\bar{C}$	D	$\bar{D}$	E	$\bar{E}$
1	1	1	2	3	2_1		3_4		3_8	3_{-3}
2		3_{-7}		1	3_5	1_{-1}		3_{-2}	2_6	2_{-6}
3	2_7		1			2_{-5}	2_2	1_{-4}	1_3	1_{-8}

It is then discovered that $2\bar{C}E\bar{D} = 1$, whence cosets 1 and 2 are the same. Examining rows 1 and 2 we find $2\bar{A} = 3$ and $1\bar{A} = 1$, so we deduce that cosets 1 and 3 are the same; thus the collapse is total. We might stop at this stage, observing that coincidence words 9 and 10 imply $B = A^{-7}$ and $B = A^4$, whence $A^{11} = I$. The programme, however, does not stop here but continues to clear up the table. Since $1\bar{B} = 3$ and $2\bar{B} = 1$, the coincidence between cosets 3 and 1 is rediscovered, no action being taken, and similarly when $1C$ and $2C$ are both found to be defined. As $1\overline{C}$ is undefined but $2\overline{C}$ is defined, the value $2\overline{C} = 1$ is transferred to $1\overline{C}$, and we get $1C = 1$, deduction 11. Similarly from $2\overline{D} = 3$ we get $1\overline{D} = 3$, deduction 12; note that the coincidence between cosets 3 and 1 has been recorded but not yet dealt with. Similarly $2E2$ becomes $1E1$, deduction 13. This deals with coset 2; we now have to eliminate coset 3. From $3B = 1$ we get $1B = 1$ and from $3D = 1$ we get $1D = 1$, deductions 14 and 15. This completes the work, and expresses all the other generators as explicit powers of A, confirming the result implied by the collapse that A generates the whole group.

This working is readily transcribed into the following conventional formal proof, in which numbers in the left margin indicate use of the numbered relation:

	$C = AB$	(1)
(1)	$D = BC = BAB$	(2)
	$E = B\overline{A}$	(3)
(3)	$D = A\overline{E} = A^2\overline{B}$	(4)
(4)	$C = \overline{B}D = \overline{B}A^2\overline{B}$	(5)
(5, 2)	$E = CD = \overline{B}A^3B$	(6)
(2, 6)	$A = DE = BA^4B$	(7)
(7)	$E = B\overline{A} = A^{-4}\overline{B}$	(8)
(1, 8, 4)	$I = \overline{C}E\overline{D} = \overline{B}A^{-7}$	(9)
(7, 9)	$I = AI\overline{A} = BA^{-4}$	(10)

This gives $A^{11} = I$ and $B = A^4$, from which the other generators can be expressed as powers of A from the cyclic relationship of the generators. Of course nothing in the programme takes advantage of any such symmetry. The lack of any machine equivalent of the concept of "similarly" is often conspicuous; sometimes the computer effectively repeats sequences of operations where human inspection would detect the similar state and move at once to the conclusion. The next example illustrates this point also.

10. The abstract group $G^{3,7,16}$

Coxeter [4] suggested that the abstract group

$$G^{3,m,n}: A^3 = B^m = C^n = (BC)^2 = (CA)^2 = (AB)^2 = (ABC)^2 = I$$

should be finite or trivial if and only if $\cos(4\pi/m) + \cos(4\pi/n) < \frac{1}{2}$, but he proved this only for cases with m, n both even. He directed attention to the case of $G^{3,7,16}$, this being the only such group not then known to be finite which satisfied his criterion. It was later proved by abstract methods [12] to be of order 21504, and this was confirmed by coset enumeration. It has a subgroup

$$(8, 7 \mid 2, 3): R^8 = S^7 = (RS)^2 = (R^{-1}S)^3 = I$$

of index 2 and so of order 10752; thus the octahedral subgroup $\{R^2, R^{-1}S\}$ has 448 cosets in this subgroup. The enumeration of these cosets, however, is far from straightforward, involving definition and subsequent elimination of many redundant cosets. This seems to be due to the difficulty of proving that the defining relations imply $(R^2S^4)^6 = I$; if this relation is added to the defining set, the enumeration presents no difficulty. I have a computer enumeration with a full printed record similar in style to those given above but without the construction and printing of deduction words. A formal proof of $(R^2S^4)^6 = I$ could be constructed from this, but I have not attempted this. Instead I followed the lines of the published abstract proof [12].

In this proof the normal subgroup generated by conjugates of R^4 is shown to be generated by seven of them (or indeed by any six of these, but it is convenient to keep seven for reasons of symmetry). Denoted by A, B, C, D, E, F, G, these are found to satisfy the relations

$$A^2 = B^2 = C^2 = D^2 = E^2 = F^2 = G^2 = I \tag{1}$$

$$(AC)^2 = (BD)^2 = (CE)^2 = (DF)^2 = (EG)^2 = (FA)^2 = (GB)^2 = I \tag{2}$$

$$(ABC)^2 = (BCD)^2 = (CDE)^2 = (DEF)^2 = (EFG)^2 = (FGA)^2 = (GAB)^2 = I \tag{3}$$

$$ABCDEFG = AFDBGEC = I \tag{4}$$

This subgroup being normal and of index 168 (the factor group is $LF(2, 7)$ in the presentation $(4, 7 \mid 2, 3)$), the problem is that of establishing its order. It turns out to be elementary Abelian of order 64, though the proof is not easy. A computer proof is given below, based on enumeration of the 16 cosets of $\{A, C\}$. Full details would take too much space as the computer output shows definition of 37 cosets before these collapse to 16, and 118 lines of working obtaining deduction words or coincidence words. The working below is presented in abstract form by extracting from the computer working the substitutions needed to prove the relations $(BE)^2 = I$ and $(AB)^2(BF)^2 = I$ which were obtained as coincidence words. Numbers in the left margin indicate that one of the relations above with that number was used to effect a substitution for the corresponding underlined part of the word in the line above (but relations (1), equating each element to its inverse, are used freely without comment).

(3)	$I = G\,F\,E\,G\,F\,E$
(2, 2)	$= B\,G\,B\,F\,G\,E\,F\,E$
(4, 3)	$= B\,E\,C\,A\,F\,D\,F\,G\,F\,D\,E\,F\,D\,F\,E$
(2, 3, 3, 2)	$= B\,E\,C\,A\,D\,E\,F\,G\,E\,C\,E\,D\,C\,D\,E$
(4, 2, 3)	$= B\,E\,C\,A\,C\,B\,A\,C\,B\,C\,D\,B\,D\,E$
(3, 2)	$= B\,E\,C\,C\,B\,E$
(1)	$= B\,E\,B\,E$

This shows that B and E commute, and we see (though the computer had to prove) that similarly any two letters commute if they are three apart in cyclic order:

(3)	$I = A\,G\,B\,A\,G\,B$
(2, 2, 4)	$= F\,A\,F\,E\,G\,E\,C\,D\,E\,F\,B$
(3, 3)	$= F\,A\,F\,D\,F\,E\,D\,F\,G\,D\,C\,F\,B$
(2, 3, 3)	$= F\,A\,D\,C\,D\,E\,C\,E\,G\,F\,E\,D\,C\,F\,B$
(3, 2, 4)	$= F\,A\,D\,B\,D\,C\,B\,C\,A\,B\,F\,B$
(2, 3)	$= F\,A\,B\,C\,A\,C\,F\,B$
(2)	$= F\,A\,B\,A\,F\,B$
(1)	$= F.A\,B\,A\,B.B\,F\,B$
whence	$(AB)^2(BF)^2 = I$

Since $(FB)^2 = I$ is a cyclic variant of $(BE)^2 = I$, we infer that $(AB)^2 = I$, i.e. A, B commute, and so by symmetry that any two adjacent letters commute, including G with A. This completes the proof that this subgroup is Abelian. The relation $(AB)^2 = I$ is equivalent to $(R^2S^4)^6 = I$.

This proof is quite unlike that given in [12], obtained empirically and tidied up for perspicuity in publication. The present proof is tidied up only by selection of the relations to be proved, and shortening the working by performing several disjoint substitutions at once instead of in separate stages. Its first part is simpler than in [12], but the second part is less so; the corresponding part of [12] would read

(4, 4)	$I = \underline{DEF}G\underline{ABC.ABC}D\underline{EFG}$
(3, 3, 3)	$= FEDGDGFE$
whence	$(FE)^2(DG)^2 = I$

So we see that computer proofs are not necessarily the shortest or simplest.

11. The abstract group $G^{3,7,17}$

As remarked in Section 10, $G^{3,7,16}$ was the only example of a group $G^{3,m,n}$ whose parameters satisfy Coxeter's suggested criterion and which was not already known to be finite. $G^{3,7,18}$ was found to be infinite [13, 10], but $G^{3,7,17}$, though not satisfying Coxeter's criterion, was still unknown. It was with some surprise that a report (unpublished, but see [5]) was received that this group collapses. This was first found at the SRC Atlas Computer Laboratory, Chilton, by J. McKay, and later confirmed in other computers. In view of the unsatisfactory state of having such a result obtained only by computers (cf. remarks in Section 1), I have performed this enumeration using the dihedral subgroup $\{AB, C\}$, with a program which prints a full record of the working, but without the construction of deduction and coincidence words. The following formal proof is extracted from this working, but rather heavily edited for perspicuity. In particular the deduction of the total collapse from the first coincidence is not based on the computer sequence. The main debt to the computer working is the identification of the critical relation which precipitates the collapse and the indicated proof of this relation. For perspicuity this proof is presented in steps leading to intermediate results. As in Section 10, the numbers in the left margin indicate the relations used in substituting for the underlined parts of the preceding expression; these may be the defining relations or relations already proved. Relations (1–7) define the group.

$$I = A^3 \tag{1}$$
$$= B^7 \tag{2}$$
$$= C^{17} \tag{3}$$
$$= (AB)^2 \tag{4}$$
$$= (BC)^2 \tag{5}$$
$$= (CA)^2 \tag{6}$$
$$= (ABC)^2 \tag{7}$$

(7) $I = \underline{A\,B}\,\underline{C\,A}\,\underline{B\,C}$

(4, 6, 5) $= \bar{B}\,\underline{\bar{A}\,\bar{A}}\,\bar{C}\,\bar{C}\,\bar{B}$

(1) $= \bar{B}\,A\,\bar{C}\,\bar{C}\,B$

i.e. $A = B^2C^2$ (8)

(8, 4) $I = C\,C\,\underline{\bar{A}}\,B\,B\,.\,B\,\underline{A}\,B\,A$

(4, 8) $= C\,C\,B\,A\,\underline{B\,B\,B\,B\,B\,B}\,C\,C\,B\,A$

(2) $= C\,C\,B\,A\,\bar{B}\,C\,C\,B\,A$

whence $B\,\bar{A}\,\bar{B} = C\,C\,B\,A\,C\,C$

and (1) $I = (C\,C\,B\,A\,C\,C)^3$ (9)

(9, 9) $I = (C^4B\,A)^3(B\,A\,C^4)^3$

(4) $= (C^4B\,A\,C^4B\,A\,C^4)^2$

$= (B\,A\,C^4B\,A\,C^8)^2$ (10)

(8) $I = C\,C\,\underline{\bar{A}}\,B\,B$

(4) $= C\,\underline{C\,B}\,\underline{A}\,\underline{B\,B\,B}$

(5, 6, 2) $= C\,\bar{B}\,\bar{C}\,\bar{C}\,\underline{\bar{A}}\,\bar{C}\,\bar{B}\,\underline{\bar{B}\,\bar{B}}\,\bar{B}$

(8, 8) $= C\,\bar{B}\,\bar{C}\,\bar{C}\,\bar{C}\,\underline{\bar{C}\,\bar{B}}\,\underline{\bar{B}\,\bar{C}\,\bar{B}}\,C\,C\,\bar{A}\,\bar{B}$

(5, 5) $= C\,\bar{B}\,\bar{C}\,\bar{C}\,\bar{C}\,B\,C\,C\,C\,C\,\bar{A}\,\bar{B}$

whence $B\,A = C\,\bar{B}\,C^{-3}B\,C^4$ (11)

(10) $I = (B\,A\,C^4\underline{B\,A}\,C^8)^2$

(11) $= (B\,A\,C^5\bar{B}\,C^{-3}B\,\underline{C^{12}})^2$

(3) $= (B\,A\,C^5\bar{B}\,C^{-3}B\,C^{-5})^2$ (12)

Note that this is the only full use of the period of C being 17. Other uses of (3) are purely negative; the period of C is odd or not divisible by 3 or 5. This relation (12), in the form

$$B\,C^{-5}B\,A\,C^5\bar{B}\,C^{-3}B\,C^{-5}\bar{A}\,\bar{B}\,C^5\bar{B} \in \{AB, C\}$$

is the summit of the coset cardhouse which precipitates its collapse. Note that

$BC^{-5}BAC^5\bar{B}$ is a conjugate of BA and so of period 2. From (12), in the form

$$B\,C^{-5}B\,A\,C^5\bar{B}\,.\,C^{-3}.\,B\,C^{-5}\bar{A}\,\bar{B}\,C^5\bar{B} = C^3$$

we obtain (3)

$$B\,C^{-5}B\,A\,C^5\bar{B}\,.\,C^{-n}.\,B\,C^{-5}\bar{A}\,\bar{B}\,C^5\bar{B} = C^n$$

and, for $n = 1$,

$$\bar{A}\,\bar{B}\,C^5\bar{B}\,\bar{C}\,B\,.\,C^{-5}.\,\bar{A}\,\bar{B}\,C^5\bar{B}\,\bar{C}\,B = C^5 \qquad (13)$$

A rather similar relation is obtained more shortly:

(6) $\qquad I = (A\,C)^2$

(8) $\qquad = (C\,C\,\underline{\bar{A}}\,B\,B\,\underline{A}\,C)^2$

(6, 8) $\qquad = (C\,C\,C\,\underline{A}\,C\,\underline{B\,B\,B\,B}\,C\,C\,C)^2$

(8, 2) $\qquad = (C\,C\,\underline{C\,B}\,B\,C\,C\,C\,\underline{\bar{B}\,\bar{B}}\,\bar{B}\,C\,C\,C)^2$

(5, 8) $\qquad = (C\,C\,\bar{B}\,\bar{C}\,B\,C\,C\,C\,C\,C\,\bar{A}\,\bar{B}\,C\,C\,C)^2$

whence $\qquad \bar{A}\,\bar{B}\,C^5\bar{B}\,\bar{C}\,B\,.\,C^5.\,\bar{A}\,\bar{B}\,C^5\bar{B}\,\bar{C}\,B = C^{-5} \qquad (14)$

Comparison of (13) and (14) leads to the triviality of the group. Let

$$W = \bar{A}\,\bar{B}\,C^5\bar{B}\,\bar{C}\,B$$

Then (13) $\qquad W\,C^{-5}\,W = C^5$

and (14) $\qquad W\,C^5\,W = C^{-5}$

whence $\qquad C^{10} = W\,C^{-5}\,W.\,\bar{W}\,C^{-5}\,\bar{W}$

$\qquad = W\,C^{-10}\,\bar{W}$

and (3) $\qquad C^n = W\,C^{-n}\,\bar{W}$

$n = 5$ gives $\qquad C^5 = W\,C^{-5}\,\bar{W}$

which, with (13) $\qquad C^5 = W\,C^{-5}\,W$

gives $\qquad W = \bar{W}$

Thus $\qquad I = W^2$

$\qquad = (\underline{\bar{A}\,\bar{B}}\,C^5\underline{\bar{B}\,\bar{C}}\,B)^2$

(4, 5) $\qquad = (B\,\underline{A}\,C^6B^2)^2$

(8) $\qquad = (\underline{B}^3C^8\underline{B}^2)^2$

(2) $= (\underline{B}^{-2}C^8)^2$

(8) $= (C^2\bar{A}C^8)^2$

(6) $= C^3A\,C^9.\,C^2\bar{A}C^8$

$= \bar{A}C^{11}A\,C^{11}$

and so $\bar{A}C^{11}A = C^{-11}$

But this is incompatible with (1, 3), so we have a total collapse to a single element.

This working is conspicuously shorter than the raw computer output, which is a wad more than half an inch thick. A substantial part of this output consists of working which plays no part in the final collapse. The last coset to be defined was numbered 1544, though the next highest numbered coset to play a part in the working was 1142. Intermediate cosets were defined by the program operating by systematically filling blanks in the coset multiplication table. It is not foreseen in a general purpose program that so many cosets thus defined will not be used.

12. The abstract group $G^{3,9,10}$

This group satisfies Coxeter's criterion (Section 10) for being finite or trivial. It has a subgroup

$$(2, 3, 9; 5): R^2 = S^3 = (RS)^9 = (\bar{R}\bar{S}RS)^5 = I$$

of index 2, which in turn has a normal subgroup of order 3 with factor group

$$((5, 5, 5; 3)): A^2 = B^2 = C^2 = (BC)^5 = (CA)^5 = (AB)^5 = (ABC)^3 = I$$

Coxeter ([3], footnote p. 143) reported that coset enumerations, carried out independently by Sinkov and himself, showed the total collapse of this last group, implying also that of the others. Later work by Sinkov [14] on representation of linear fractional groups established that $LF(2, 19)$ has a representation as a factor group of (2, 3, 9; 5), showing that this cannot be trivial. Pursuing this, he performed computer coset enumerations, showing that ((5, 5, 5; 3)) is $LF(2, 19)$ and that (2, 3, 9; 5) is the direct product of this with the cyclic group C_3. I have confirmed these results, finding that in each case many redundant cosets are defined before a partial collapse reduces this to the final number.

I have examined the enumeration for ((5, 5, 5; 3)) in some detail. I used the icosahedral subgroup $\{AB, C\}$, which is of index 57, and found that the computer defined 226 cosets before coincidences were found. More than 20 of these cosets play no part in the working, but it is clear that many more than 57 cosets are essentially involved in the working, an overshoot and collapse being unavoidable. In hand work it is extremely easy to mishandle such a partial collapse, and no doubt this is what Sinkov and Coxeter did in deducing a total collapse (I have had a sporadic computer malfunction giving a total collapse on this example).

In this example the difficulty seems to be that the relations ((5, 5, 5; 3)) imply relations such as $(B\,C\,A\,B\,A\,B\,C\,B\,A\,B\,A\,C)^2 = I$, i.e. B commutes with $C(AB)^2C(AB)^{-2}C$, and variations of this under permutations of A, B, C, but that there is no short proof of this implication. (It is much easier to prove that these variant relations are equivalent than to prove any one of them.) The relation was discovered by noticing that the coincidence which involved the smallest coset numbers implied that $B\,C\,A\,B\,A\,B\,C\,B\,A\,B\,A\,C\,B \in \{AB, C\}$, and that the coincidence word, appearing in a long and complicated form, was reducible to $C\,A\,B\,A\,B\,C\,B\,A\,B\,A\,C$. However, although the computer working constitutes, and indeed exhibits, a proof of this relation, I have not succeeded in simplifying this to a version fit for public exhibition. If this relation is adjoined to the defining relations, the enumeration presents no difficulties, and can be performed (though not with the computer sequence of definitions) without redundant cosets.

13. Reidemeister-Schreier relations

When a coset enumeration has shown that certain elements generate a subgroup of finite index in a finitely presented group, the work does not exhibit defining relations for the subgroup. The following adaptation enables such relations to be deduced. In Section 8 the mode of construction of coincidence words was given. Two ways, very similar to these, of constructing "non-coincidence words" are as follows.

First, suppose that the product of a coset by a relator is found to be the same coset, so that no coincidence results. Hitherto we have taken no further action and continued with the working. Now we treat the relator analogously to one giving a coincidence, substituting for any letter corresponding to a deduction the deduction word corresponding to it. On cancelling as much as possible, we find either that this word becomes empty or that it is a conjugate of a non-empty relator in the subgroup generators. In the latter case we print the word as a non-coincidence word.

Second, suppose that we find an apparently new coincidence, but discover, as described in Section 7, that it gives no new information, leading only to a "coincidence" between a coset and itself. Then the "coincidence word", constructed as in Section 8 when placing a new coincidence in the list, will again (if not empty) be a conjugate of a relator in the subgroup generators. This too is printed out.

The relators obtained in this way give, usually with considerable redundancy, a defining set of relators for the subgroup. Further analysis may allow the set to be reduced.

An example is taken from Section 9. The coincidence 3 = 1 was discovered from 2 = 1 on finding that $2\bar{A} = 3$ and $1\bar{A} = 1$, with coincidence word BA^{-4}. It is rediscovered on finding that $1\bar{B} = 3$ and $2\bar{B} = 1$, with coincidence word $3B1\bar{I}2\bar{B}1 = BA^7B\bar{B} = BA^7$. When placing this in the coincidence list we find it there already, and deduce the non-coincidence word $A^{-7}\bar{B}BA^{-4} = A^{-11}$, proving that $A^{11} = I$ is a relation for the subgroup. From $1C = 2$ and $2C = 3$ we rediscover this relation, and it also turns up several times when completing the ordinary working, such as $1I1 = AB\bar{C} = A\,.\,A^4\,.\,A^6 = A^{11}$. Since no other power appears, this is the sole defining relation for the subgroup, which is in fact the whole group.

14. A non-Hopfian group

Baumslag and Solitar [1] have given examples of non-Hopfian groups with two generators and one relation, the relation having the form

$$A^{-1}B^mA = B^n$$

The simplest non-Hopfian group of this type has the relation

$$A^{-1}B^2A = B^3$$

The elements A and $C = B^{2^n}$ also generate this group, and satisfy the relation

$$A^{-1}C^2A = C^3$$

but this relation is no longer sufficient to define the group completely. I have examined the cases $C = B^2$ and $D = B^4$, using the technique of Section 13 to obtain relations. For $C = B^2$, after defining the subgroup $\{A, B^2\}$ by $1A = 1$, $1B = 2$, $2B = 1$, I obtained the following working. No further cosets were defined.

1 2 = 1 BY 1 FROM $\bar{A}BBA\bar{B}\bar{B}\bar{B}$

$BBB\bar{A}\bar{B}\bar{B}A$

2 1 = 1

$BB\bar{A}\bar{B}\bar{B}ABBBB\bar{A}\bar{B}\bar{B}A$

3 1 B 1

$\bar{A}BBA\bar{B}\bar{B}$

4   1 = 1   BY   1   FROM $BBA\bar{B}\bar{B}\bar{B}\bar{A}$

$\bar{A}BBA\bar{B}\bar{B}\bar{A}BBA\bar{B}\bar{B}ABB\bar{A}\bar{B}\bar{B}ABB\bar{A}\bar{B}\bar{B}ABB\bar{A}\bar{B}\bar{B}$

Of these, lines 2 and 4 give non-coincidence words as described in Section 13. (The program also rediscovers the conclusion 4 in cyclic variant forms by applying $1B1$ to each of the occurrences of B in the relator $\bar{A}BBA\bar{B}\bar{B}\bar{B}$. I do not reproduce these here.)

In terms of the generators A and $C = B^2$, the group is thus found to satisfy the relations

$$C\bar{A}\bar{C}ACC\bar{A}\bar{C}A = I \tag{2}$$

and

$$\bar{A}CA\bar{C}\bar{A}CA\bar{C}A\underline{C\bar{A}\bar{C}AC\bar{A}\bar{C}AC}\bar{A}\bar{C} = I \tag{4}$$

The latter relation is a consequence of the former, as we can see by noticing that the underlined letters form a cyclic permutation of (2) and that after cancelling these and the letters $A, \bar{A}$ thereby made adjacent, we are left with the inverse of a cyclic permutation of (2).

It is worth observing that the relation $\bar{A}C^2A = C^3$, apparently absent from these relations, is implied by (2), as we may see thus. (2) is equivalent to $C = (\bar{A}CA\bar{C})^2$, so C commutes with $\bar{A}CA\bar{C}$ and so with $\bar{A}CA$. Thus

$$\begin{aligned} I &= \bar{C}\bar{C}\bar{A}\underline{CA\bar{C}}\bar{A}CA \\ &= \bar{C}\bar{C}\bar{C}\bar{A}CA\bar{A}CA \\ &= C^{-3}\bar{A}C^2A \end{aligned}$$

as required. But $\bar{A}C^2A = C^3$ does not itself imply (2).

In the second case, with the generators A, $D = B^4$, the subgroup is defined by setting $1A = 1$, $1B = 2$, $2B = 3$, $3B = 4$, $4B = 1$. No further cosets are defined, and the following working is obtained. I simplify the deduction and other words by writing D for $BBBB$ and $\bar{D}$ for $\bar{B}\bar{B}\bar{B}\bar{B}$ wherever these occur.

1   3 A 4   BY   1   FROM $\bar{A}BBA\bar{B}\bar{B}\bar{B}$

$\bar{B}\bar{B}ABBB$

2   4 = 2   BY   1   FROM $A\bar{B}\bar{B}\bar{B}\bar{A}BB$

$\bar{B}\bar{B}\bar{B}\bar{A}DA\bar{B}\bar{B}\bar{B}$

3   3 A 2

$BBA\bar{B}\bar{B}\bar{B}$

4 3 = 1

$BB\bar{A}\bar{D}AD$

5 1 = 1

$\bar{D}\bar{A}DA\bar{D}\bar{A}DA\bar{D}$

6 2 = 1

$BBB\bar{A}\bar{A}\bar{D}ADA$

7 2 *B* 1

$BBB\bar{A}\bar{D}AD$

8 1 = 1

$\bar{A}\bar{D}\bar{A}DAA\bar{D}\bar{A}\bar{D}\bar{A}DA\bar{D}AD$

9 1 *B* 1

$D\bar{A}\bar{A}\bar{D}ADA$

10 1 = 1 BY 1 FROM $BBA\bar{B}\bar{B}\bar{B}\bar{A}$

$D\bar{A}\bar{A}\bar{D}ADAD\bar{A}\bar{A}\bar{D}ADA\bar{D}\bar{A}DAA\bar{D}\bar{A}\bar{D}\bar{A}DAA\bar{D}\bar{A}\bar{D}\bar{A}DAA\bar{D}\bar{A}$

Of these, lines 5, 8 and 10 give the non-coincidence words forming relators in the generators *A*, *D*. (Again the last is obtained in cyclic variant forms not reproduced here.)

We thus have the relations

$$\bar{D}\bar{A}DA\bar{D}\bar{A}DA\bar{D} = I \tag{5}$$

$$\bar{A}\bar{D}\bar{A}DAA\bar{D}\bar{A}\bar{D}\bar{A}DA\bar{D}AD = I \tag{8}$$

$$D\underline{\bar{A}\bar{A}\bar{D}ADAD\bar{A}\bar{A}\bar{D}ADA\bar{D}\bar{A}D}AA\bar{D}\underline{\bar{A}\bar{D}\bar{A}DAA\bar{D}\bar{A}\bar{D}\bar{A}DA}A\bar{D}\bar{A} = I \tag{10}$$

The last is a consequence of the others, as we can see by making two applications of (8). The first underlined sequence is inverse to a cyclic permutation of (8), while the other is an almost complete sequence, being equal to $\bar{D}\bar{A}D$ by (8). These changes being made, (10) becomes $DA\bar{D}\bar{D}\bar{A}DAD\bar{A} = I$, a cyclic permutation of (5).

We have thus reduced the relations to two. These are independent, for *G*. Higman (quoted in [1]) has shown that in terms of A and B^4 there cannot be a one-relation presentation. (5) has the same form as (2) for A, C, and thus implies similarly that $\bar{A}D^2A = D^3$ and that D commutes with $\bar{A}DA$ and $AD\bar{A}$. Using these, we can simplify (8); thus

$$
\begin{aligned}
&(8) & I &= AA\bar{D}\bar{A}\underline{\bar{D}\bar{A}DA}\bar{D}AD\bar{A}\bar{D}\bar{A}D \\
&(2) & &= AA\bar{D}\bar{A}\underline{D\bar{A}\bar{D}A}\,\underline{AD\bar{A}\bar{D}}\bar{A}D \\
&(\text{commutation}) & &= AA\bar{D}\bar{A}\bar{A}\bar{D}AD\bar{D}AD\bar{A}\bar{A}D \\
& & &= A^2\bar{D}A^{-2}\bar{D}A^2DA^{-2}D
\end{aligned}
$$

and so D commutes with A^2DA^{-2}.

15. Conclusion

In its simplest form, coset enumeration gives the index of a subgroup in a finitely presented group with almost no information to supplement this bare fact. In the present work the object has been to supplement this information in various ways. Implied proofs of relations such as those giving periods of elements are obtained as formal proofs. In cases of collapse, formal proofs that the subgroup is the whole group are obtained, which may imply the collapse of the whole group. Finally, relations for the subgroup are obtained from the enumerations.

Rather frequently, especially in the more complicated cases involving many cosets, the formal proofs are long and far from perspicuous. Sometimes human editing can ease this and lead to perspicuous proofs. On other occasions it is likely that no such comprehensible proof is possible. Even then, however, one's confidence in the computer's results may be enhanced by the exhibition of its proofs, though they are scarcely more penetrable than the bare results.

The work is necessarily restricted to particular groups, rather than giving results about classes of groups. With groups, however, as with people, there is plenty of scope for the study of individuals as well as of classes.

References

1. G. Baumslag and D. Solitar. Some two-generator one-relator non-Hopfian groups. *Bull. Amer. Math. Soc.* **68**(1962), 199–201.
2. J. J. Cannon, L. A. Dimino, G. Havas, and J. M. Watson. Implementation and analysis of the Todd-Coxeter algorithm. *Mathematics of Computation* 27(1973), 463–490.
3. H. S. M. Coxeter. The abstract groups $G^{m,n,p}$. *Trans. Amer. Math. Soc.* **45** (1939), 73–150.

4. H. S. M. Coxeter. Groups generated by unitary reflections of period 2. *Can. J. Math.* **9**(1957), 243–272.
5. H. S. M. Coxeter. "Twisted Honeycombs". American Mathematical Society, 1970.
6. H. S. M. Coxeter and W. O. J. Moser. "Generators and Relations for Discrete Groups". Springer-Verlag, 1957, 1965, 1972.
7. J. Leech. Some definitions of Klein's simple group of order 168 and other groups. *Proc. Glasgow Math. Assoc.* **5**(1962), 166–175.
8. J. Leech. Coset enumeration on digital computers. *Proc. Cambridge Phil. Soc.* **59**(1963), 257–267.
9. J. Leech. Generators for certain normal subgroups of (2, 3, 7). *Proc. Cambridge Phil. Soc.* **61**(1965), 321–332.
10. J. Leech. Note on the abstract group (2, 3, 7; 9). *Proc. Cambridge Phil. Soc.* **62**(1966), 7–10.
11. J. Leech (Ed.). "Computational Problems in Abstract Algebra". Pergamon, 1970 (especially J. Leech: Coset enumeration, pp. 21–35).
12. J. Leech and J. Mennicke. Note on a conjecture of Coxeter. *Proc. Glasgow Math. Assoc.* **5**(1961), 25–29.
13. C. C. Sims. On the group (2, 3, 7; 9). *Notices Amer. Math. Soc.* **11**(1964), 687–688.
14. A. Sinkov. The number of abstract definitions of $LF(2, p)$ as a quotient group of (2, 3, n). *J. Algebra* **12**(1969), 525–532.
15. J. A. Todd and H. S. M. Coxeter. A practical method for enumerating cosets of a finite abstract group. *Proc. Edinburgh Math. Soc.* (2) **5**(1936), 26–34.

Notes added in proof

Groups of the same type as that in Section 9, called *Fibonacci groups*, have been the subject of several recent papers. In particular, Havas [18] has given a computer-derived proof that the Fibonacci group on seven generators is the cyclic group of order 29. Brunner [17] has given a detailed account of the non-Hopfian group of Section 14, and Beetham and Campbell [16] have made computer studies of this group.

16. M. J. Beetham and C. M. Campbell. A note on the Todd-Coxeter algorithm. *Proc. Edinburgh Math. Soc.* (2) **20**(1976), 73–79.
17. A. M. Brunner. Transitivity systems of certain one-relator groups. *Proc. Second Internat. Conf. Th. Groups*, Lecture Notes in Mathematics (Springer, Berlin) **372**(1974), 131–140.
18. G. Havas. Computer aided determination of a Fibonacci group. *Bull. Austral. Math. Soc.* **15**(1976), 297–305.

3. The Miracle Octad Generator

J. H. CONWAY

Department of Pure Mathematics, University of Cambridge, Cambridge, England

The Miracle Octad Generator (MOG) is a device invented by Robert Curtis for computing with the octads of the Steiner system $S(5, 8, 24)$ and the permutations of the Mathieu group M_{24}. In my lectures at the conference I explained the construction of the MOG in some detail, and used it to give a fairly complete discussion of the subgroups of M_{24}. I also made some remarks about subgroups of .0. However, the maximal subgroups of M_{24} are discussed (not using the MOG) in [1] and (using it) in [2], and in [3] Curtis uses the MOG to find all subgroups of .0 that fix some non-zero vector of the Leech lattice. So in this note I merely define the MOG and show how to use it to find octads of $S(5, 8, 24)$ from five of their points, and how to find permutations of M_{24} satisfying some simple conditions.

Let Ω be the set $\{\infty, 0, 1, 2, \ldots, 22\}$ and α be the permutation $x \to x + 1$ (mod 23) of these. Let Q be the set $\{0, 1, 2, 4, 8, 16, 9, 18, 13, 3, 6, 12\}$ consisting of 0 and the quadratic residues (or powers of 2), and N the complementary set. By applying α, N yields 23 12-element sets, N, N_1, N_2, etc. where N_i contains the numbers of the form $n + i$ $(n \in N)$. From these by taking symmetric differences we get 2^{12} sets in all, which we call $\mathscr{C}$*-sets.* If we replace each set by its characteristic function (interpreted in $GF(2)$), then the $\mathscr{C}$*-sets* correspond to the vectors of the *extended Golay code.*

Since $N + N_1 + N_2 + \ldots = \Omega$ (we use + for symmetric difference of sets), Ω is itself a $\mathscr{C}$-set (and therefore so is Q), and the $\mathscr{C}$-sets come in complementary pairs. In fact the $\mathscr{C}$-sets comprise the empty set $\emptyset$ and its complement Ω, 2576 12-element $\mathscr{C}$-sets called (*umbral*) *dodecads,* and 759 8-element $\mathscr{C}$-sets called (*special*) *octads* and their complementary 16-element sets. Any 5 points of Ω are contained in just one octad, or in other words, the octads constitute a *Steiner system* $S(5, 8, 24)$. It was shown by Witt that any two systems $S(5, 8, 24)$ differ only by renaming the points, so that the system we have defined is essentially the only possible one. The Mathieu group M_{24} is the group of all permutations of Ω that take every octad to an octad.

1. How to find octads of $S(5, 8, 24)$

Figure 1 is the MOG. It has 36 *pictures*. One of these shows the members of Ω arranged in a particular order as three particular octads, which we call the *bricks*. Each other picture separates the 24 points into a brick and its complementary *square* of 16 points, and then further partitions the brick into two tetrads (*black* and *white*), and the square into four tetrads (*black, white, circle,* and *dot*). Collectively, therefore, the picture defines a partition of Ω into six tetrads – for the bottom right-hand picture these are:

$$\{\infty, 14, 0, 20\}, \{8, 3, 15, 18\}, \{17, 19, 10, 6\}, \{11, 22, 2, 21\},$$
$$\{13, 1, 7, 12\}, \{4, 9, 16, 5\}$$

Now these tetrads form what was called in [1] a *sextet,* that is to say, that the union of any two of them is an octad. So for instance

$$\{\infty, 14, 0, 20, 13, 1, 7, 12\}$$

is an octad. In a similar way, the top right-hand picture shows that $\{14, 0, 3, 20, 17, 4, 5, 6\}$ is an octad. But we can also use the 35 pictures more generally so that the separated brick represents any one of the three original bricks, and its complementary square the other two. So for instance using the brick of the bottom left-hand picture to denote the central brick of the three, and its complementary square the other two (in order) we find that the all-black octad of that picture represents the octad $\{17, 4, 16, 2, \infty, 14, 21, 6\}$.

Now the remarkable fact that makes the MOG so useful is that any octad other than one of the three bricks meets at least one of these bricks in just 4 points. On the other hand, the octads obtainable from the sextets in the 35 pictures of the MOG include all those which meet its separated brick in 4 points. So, since we allow this brick to be any one of the three, we see that:

Every octad of $S(5, 8, 24)$ can be seen as the union of 2 tetrads from some picture

Recognising given 8-element sets as octads now becomes child's play. Thus the first octad in Todd's list [4] is $\{\infty, 0, 1, 2, 3, 5, 14, 17\}$, which is the all-black octad represented by the third picture in the second row, the special brick being the leftmost one of the three. Again, the last octad of Todd's list is $\{10, 11, 12, 13, 14, 17, 20, 22\}$. If these points are marked in the MOG order, we see that the central brick contains four of them, and must therefore be the separated brick in the picture sought. This is now easily located as the last picture in the next-to-last row, and the given octad is the union of the black points of the separated brick and the white ones of its complementary square (see Fig. 2).

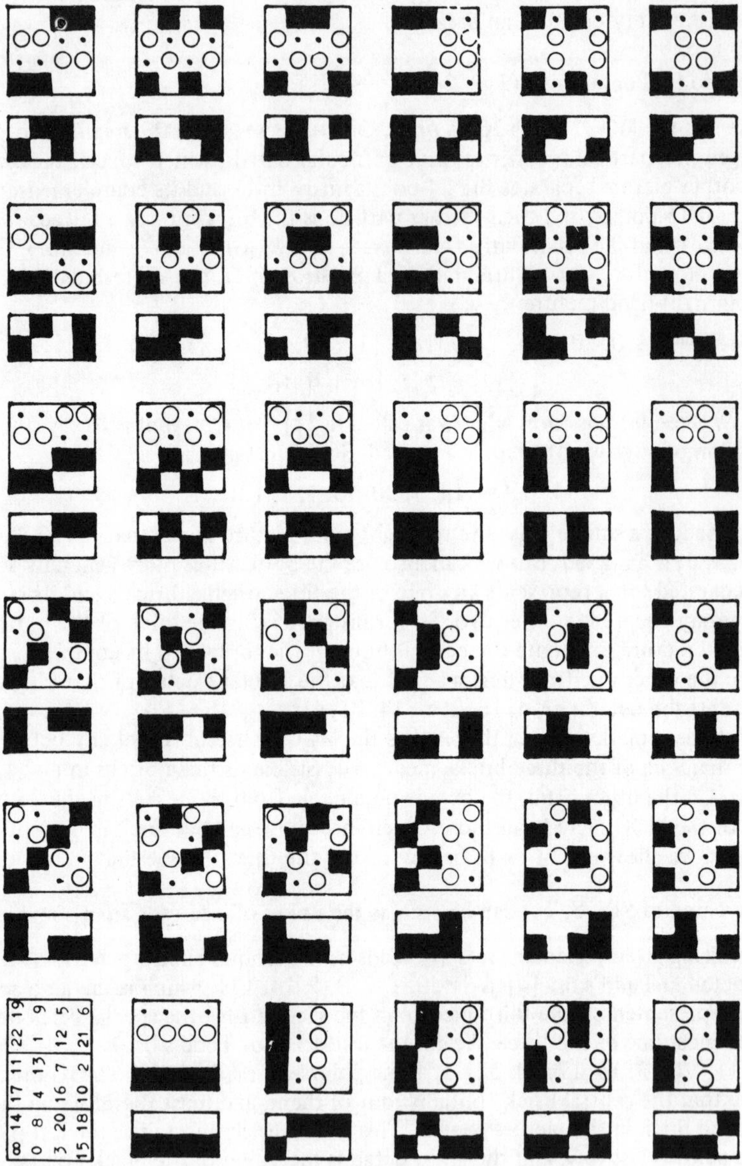

Fig. 1 The Miracle Octad Generator

```
- x|x x|x -
- -|- x|- -
- x|- -|x -
- -|x -|- -
```

Fig. 2 Recognising the octad $\{10, 11, 12, 13, 14, 17, 20, 22\}$.

2. Finding octads from five given points

This is slightly harder than verifying that a given 8-element set is indeed an octad. We first examine the tetrads that appear in the complementary squares of the MOG pictures. Close inspection reveals that any such tetrad meets each of the four rows evenly *or* each of these rows oddly, and that the corresponding statement is true for the columns. This property makes it very easy to complete any such tetrad from three of its points. Thus the tetrad defined by $\{16, 13, 21\}$ must hit both rows and columns oddly, since this triad already hits three rows and three columns, and so its fourth point must be 19, the intersection of the unused row and column. Again, the triad $\{1, 2, 4\}$ must extend to a tetrad hitting the rows evenly but the columns oddly (why?), and its fourth point is therefore 6, which is the intersection of the unused column with the presently even row.

After very little time, one gets to know these special tetrads like old fiends (the misprint here seemed too apt to be corrected). Most of them have rather nice shapes in the picture. We use them in locating octads from five given points as follows. Mark the five given points in the MOG arrangement, and examine how they split across the three bricks. Five special cases are shown in Fig. 3.

```
- -|x o|- -    - -|- -|o -    - -|- -|x x    o o|- x|- x    o -|x -|x -
- -|x x|- -    x -|- x|- -    - -|- -|- -    x -|- -|- -    - -|- -|- -
- -|x o|- -    - o|- x|- -    - x|- o|- o    - -|- x|- x    x -|- x|- o
- -|o x|- -    - -|x x|- o    - o|- x|x -    - o|- -|- -    o x|- -|- -
   (i)            (ii)            (iii)          (iv)            (v)
```

Fig. 3 Completing octads (xxxxxooo) from five given points (xxxxx).

If some brick has all five points, that brick is the desired octad (case (i)). If not, but some brick has four points, we find some picture in which these four points are coloured in the same way in the separated brick, and then the octad is found by locating the corresponding special tetrad containing the fifth point. (This happens in case (ii).) Otherwise, if some brick has three of the given points, then there are *five* pictures for which these three points are similarly coloured in the separated octad, and for just *one* of these the two remaining points are the same colour in the complementary square (case (iii)), locating the desired octad. Although this case is fairly easy to describe, such octads are the hardest to find in practice. In all other cases the given five points are split 2:2:1 across the three bricks, and there are three possible choices for the separated brick. This will be the brick containing just one of the five points if the remaining four points form a special tetrad in the complementary square. (This happens in case (iv).) If not, the separated brick is one of the two others, and for each possibility we have three of the given points in the complementary square. Complete these three to a special tetrad, and locate the corresponding MOG picture. Then for just one of the two possibilities the two remaining points will be similarly coloured in the separated brick, and the octad is located (case (v)).

In the figures the five given points are marked x, and the three points that are found by this process to complete them to an octad are marked o. Although our explanation of the 2:2:1 case is quite long, the process is very easy in practice, and such octads are usually easier to find than the 3:2:0 and 3:1:1 types. The reader is recommended to ask a friend (or fiend!) to select various sets of five points of Ω, and then to try to complete these to octads. Of course, when an octad has been found by any mixture whatever of intuition and intellect, it is trivial to verify its correctness.

3. Using the MOG for other purposes

(i) *Completing sextets.* Any four points of Ω form what is called a *tetrad*, and any tetrad belongs to a set of six mutually disjoint tetrads with the property that the union of any two is an octad (a *sextet*). Given one or more of the tetrads from a sextet, we can find others by finding the octads defined by 5-element sets made of one of the given tetrads and arbitrarily chosen additional points. By judicious choice of the latter, sextet-completion becomes easier than octad-location.

(ii) *Other $\mathscr{C}$-sets.* Of course, 16-element $\mathscr{C}$-sets satisfying given conditions are best found by locating their complementary octads. Dodecads are usually easily found as symmetric differences of octads.

(iii) *Elements of M_{24} of cycle-shape $1^6 3^6$.* Given the six fixed points, the triples of points forming the 3-cycles are located as the completions of the six sets of five fixed points to octads. The senses of rotation can then be correlated by observing the effect of the desired permutation on randomly chosen octads. (The six fixed points of such elements cannot lie in the same octad.)

(iv) *Elements of M_{24} of cycle-shape $1^8 2^8$.* Here the fixed points form an octad. Given the fixed points, there is just one element of this kind that interchanges any two points from the complementary 16-element set, say the points a and b. Then if we find any octad made of *four* of the fixed points, the *two* points a and b, and *two* further points c and d, then c and d will be interchanged by the desired permutation. The cycles (ab), (cd), . . . found in this way complete the permutation.

(v) *Other elements of M_{24}.* There are similar rules for finding elements of other

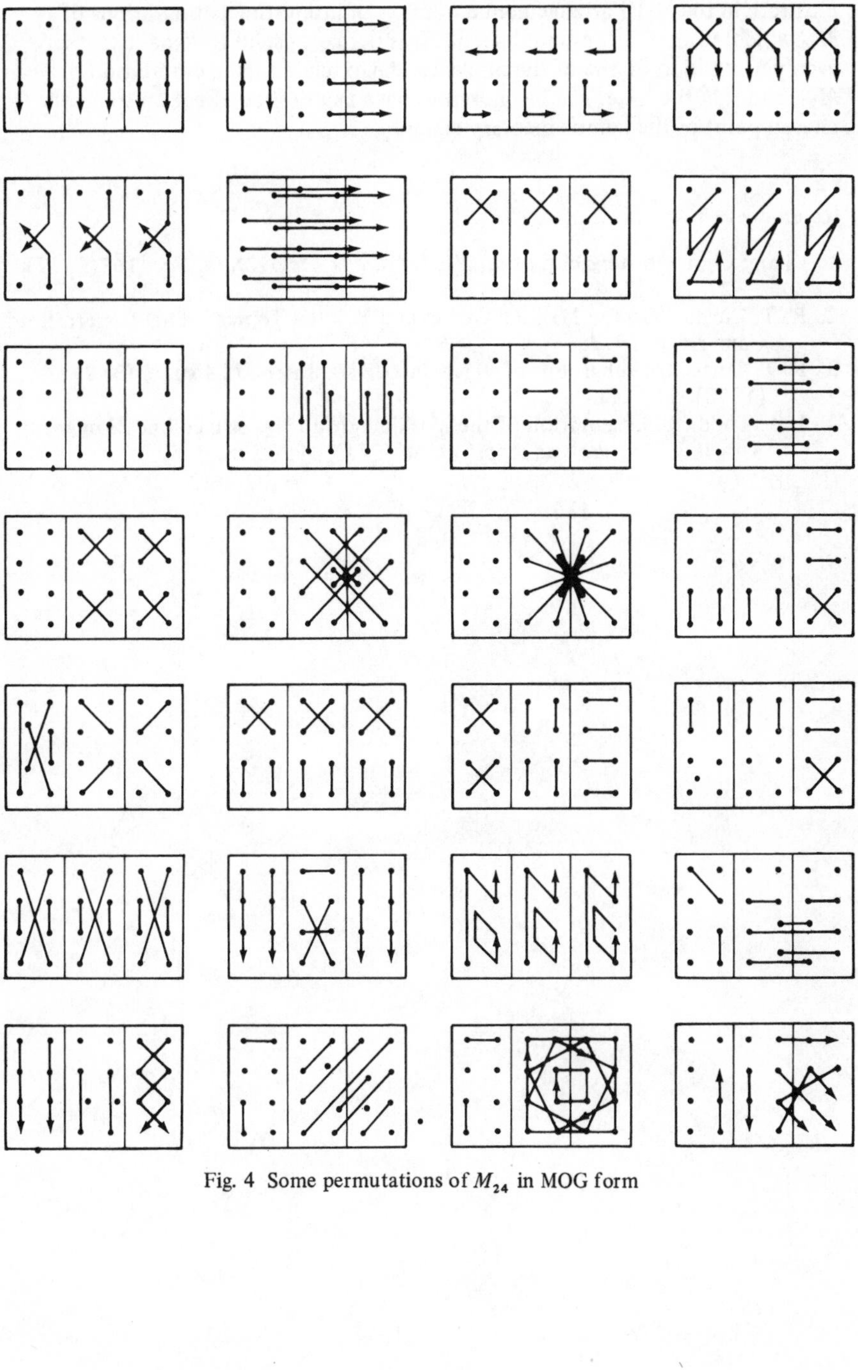

Fig. 4 Some permutations of M_{24} in MOG form

cycle-shapes which we have not found it worthwhile to give. But in Fig. 4 we illustrate in the MOG arrangement a number of interesting permutations of M_{24} which we have often found useful. It is perhaps useful to remark that any even permutation of one of the bricks can be extended to a permutation of M_{24}, and that the latter can be chosen to have any desired effect on some one chosen point of the complementary square.

References

1. Chang Choi. On subgroup of M_{24} (I, II). *Trans Amer. Math. Soc.* **167**(1972), 1–27, 29–47.
2. R. T. Curtis. "On the Mathieu Group and Related Topics". PhD Dissertation, Cambridge, 1972.
3. R. T. Curtis. On subgroups of .0 (I): lattice stabilisers. *J. Algebra* (3) 27 (1973), 549–573.
4. J. A. Todd. Representation of the Mathieu group M_{24} as a collineation group. *Annali di Mat. Pure ed Applicata* **4**(1971), 199–238.

4. A Quaternionic Construction for the Rudvalis Group

J. H. CONWAY

Department of Pure Mathematics, University of Cambridge, Cambridge, England

1. "Historical remarks"

In [1] David Wales and I constructed the simple group R of order 145,926,144,000 predicted by Arunas Rudvalis [2]. We showed that a double cover $2R$ of R was realised as a subgroup of index 2 in the automorphism group of a certain configuration of 4×4060 vectors in 28-dimensional complex space. These vectors, which we familiarly called the *sacred vectors,* fall naturally into 4060 sets of four, since whenever v is a sacred vector, so are $-v$, iv, and $-iv$. The full automorphism group of the configuration they form is a group $4R$ which has a cyclic centre of order 4 generated by i. We also loosely call $\{ v, -v, iv, -iv \}$ a vector, since of course it corresponds to a point in projective 27-space, and then R is realised as the group of collineations of the 4060 points obtained in this way from the sacred vectors as properly defined.

The construction in [1] was rather complicated†, and Wales and I spent some time trying to find simpler coordinate systems without much success. But when I showed the results of some of our efforts to Michael Guy in Cambridge he rapidly found a very natural coordinate frame, in which the coordinates of all the sacred vectors are chosen from the population $0, i^n, 4i^n$, and which displays at once many combinatorial properties of the group. Two sacred vectors not multiples of each other have scalar product 0 or $4i^n$.

Guy used a computer to study the 290 orbits of an element of order 14 from $4R$ on the sacred vectors, and found the inner products of each vector with the other vectors of its orbit. He discovered that there were just four orbits A, B, C,

† Probably because of time pressure – Rudvalis tells me that the computer output which first led him to predict his group emerged at roughly 3 pm on May 4 1972. The final calculations by which Wales and I verified its existence were completed at exactly 4 pm on June 3!

D that consisted of 14 mutually perpendicular vectors. Investigating further, he found that the vectors of A were perpendicular to those of B, and those of C were perpendicular to those of D. So we have the surprising result that from the 4060 sacred vectors we can select 28 particular ones (those of $A \cup B$, or of $C \cup D$) that form a natural coordinate frame. Guy then had the computer write all the sacred vectors in this frame, and a careful study of their coordinates led to the present paper.

In what follows we outline an approach that might have suggested the existence of such a frame quite naturally, and we do not mention Guy's results again. But it should be stressed that of course the quaternionic approach was only discovered as a result of Guy's work.

2. The quaternionic base

Let p be an element of order 7 in $2R$. Then the centraliser of p has the form $\langle p \rangle \times G$, where G is a quaternion group of order 8, whose central element is the scalar multiplication by -1, and which acts without fixed points on the 28-dimensional space. We take G to consist of the elements

$$1, \quad -1, \quad J, \quad -J, \quad K, \quad -K, \quad L, \quad -L$$

with the usual relations (but with J, K, L for the more usual i, j, k)

$$JK = L, \quad KL = J, \quad LJ = K, \quad KJ = -L, \quad LK = -J, \quad JL = -K,$$
$$J^2 = K^2 = L^2 = -1$$

Now for a suitable unit vector e_0, if we define $e_n = e_0 p^n$, the 28 vectors

$$e_0, \ldots, e_6, e_0J, \ldots, e_6J, e_0K, \ldots, e_6K, e_0L, \ldots, e_6L$$

form an orthonormal coordinate frame. This much follows easily from the action of the group $\langle p \rangle \times Q$ on the 28-space, which can be found from the character table of $2R$. What is remarkable is the fact that, in a suitable scale, we can so choose this base that the vectors $4e_n$ are sacred vectors. We shall suppose this done from now on.

Taking the typical complex quaternion $q = a + bj + ck + dl$, we define

$$e_nq = ae_n + be_nJ + ce_nK + de_nL$$

so that the general vector can be written in the form

$$v = e_0q_0 + e_1q_1 + \ldots + e_6q_6$$

and our 28-dimensional complex space becomes a 7-dimensional space with complex quaternion coordinates. This representation is not invariant under the entire group, and so has no great significance for the representation theory of R, but it reveals a number of striking symmetries, and enables us to write many of the group operations in particularly simple form.

3. The sacred vectors in quaternionic form

Table 1 shows (in column form) the quaternionic coordinates of a number of sacred vectors. Now from any such vector (q_n) we can obtain a number of others by fairly simple operations, namely

(i) cyclically permuting the q_n,
(ii) changing the signs of q_n, q_{n+3}, q_{n+5}, q_{n+6},
(iii) postmultiplying all q_n by j, k, or l.

The first listed vector yields 28 vectors in this way, the next four yield 112 each, and the last 16 give 224 each, so that we obtain in all $4060 = 28 + 4 \times 112 + 16 \times 224$ sacred vectors in this way. These include one representative from each set $\{v, -v, iv, -iv\}$.

The vectors obtained from the first vector listed form what we call the *small orbit* of 4×28 vectors, which essentially are the vectors of our natural coordinate frame. The 4×448 vectors obtained from the next four listed vectors form the *middle orbit*, and are the members of coordinate frames sharing 12 elements with our given one. Finally, the 4×3584 vectors obtained from the last 16 listed ones form the *large orbit*. These three sets are in fact orbits under the monomial group defined in the next section.

In Table 2 we give the vectors of the middle orbit in more detail, since they arise in many calculations. All middle orbit vectors (to within scalar multiplication) are found by rotations and sign-changes from those listed.

4. Some group elements

We call a group element *monomial* if it takes each vector from the small orbit to another from that orbit. The collection of all such elements in the *monomial group*, a subgroup of $4R$ which has order $2^{16} . 3 . 7$. It turns out that if a monomial element takes e_n to $e_m u$, where u is one of the 16 *quaternion units* ± 1, $\pm i$, $\pm j$, $\pm ij$, $\pm k$, $\pm ik$, $\pm l$, $\pm il$, then it takes every vector $e_n u'$ to a vector $e_m u''$ for the same m, where u' and u'' represent further quaternion units. From this remark it follows that any such element takes the typical vector (q_n) to (q'_n), where each q'_m depends on only one of the q_n. So in this sense, these elements can also be written as monomial elements in 7 coordinates.

Table 3 describes a number of group elements by tabulating the values of the q'_n in terms of the q_n. All but one of the elements in this table are monomial, and that one can be written very nicely as a 7×7 matrix of quaternion left-multiplications. Collectively, the elements from Table 3 generate the entire group $4R$.

TABLE 1

Quaternionic Coordinates for the Sacred Vectors

small orbit	middle orbit	large orbit			
	$1+j+k+l$	S	S	S	S
	0	$i-l$	$1-il$	$-j+ik$	$ij-k$
	0	$i-j$	$1-ij$	$-k+il$	$ik-l$
	$i+j+ik-l$	$-k+l$	$-ik-il$	$-i+ij$	$-1-j$
	0	$i-k$	$1-ik$	$ij-l$	$-j+il$
	$i+ij-k+l$	$-j+k$	$-ij-ik$	$-i+il$	$-1-l$
	$i-j+k+il$	$j-l$	$-ij-il$	$-i+ik$	$-1-k$
	$1+j-k-l$	S^j	S^j	S^j	S^j
	0	$j+ik$	$-ij-k$	$i+l$	$1+il$
	0	$ik+l$	$-k-il$	$1+ij$	$i+j$
	$i-j+ik+l$	$-i-ij$	$-1+j$	$k+l$	$ik-il$
	0	$1-ik$	$-i+k$	$-j+il$	$-ij+l$
4	$i-ij+k+l$	$-ij-ik$	$j-k$	$-1-l$	$i-il$
0	$i-j-k-il$	$1-k$	$-i-ik$	$ij-il$	$j+l$
0					
0	$1-j+k-l$	S^k	S^k	S^k	S^k
0	0	$1-il$	$-i+l$	$ij-k$	$j-ik$
0	0	$k+il$	$-ik-l$	$i+j$	$1+ij$
0	$i+j-ik+l$	$-ik-il$	$k-l$	$-1-j$	$i-ij$
	0	$j+il$	$-ij-l$	$1+ik$	$i+k$
	$i-ij-k-l$	$1-l$	$-i-il$	$-ij+ik$	$j+k$
	$i+j-k+il$	$-i-ik$	$-1+k$	$j+l$	$-ij+il$
	$1-j-k+l$	S^l	S^l	S^l	S^l
	0	$ij+k$	$-j-ik$	$1+il$	$i+l$
	0	$1-ij$	$-i+j$	$ik-l$	$k-il$
	$i-j-ik-l$	$1-j$	$-i-ij$	$-ik+il$	$k+l$
	0	$ij+l$	$-j-il$	$i+k$	$1+ik$
	$i+ij+k-l$	$-i-il$	$-1+l$	$j+k$	$ij-ik$
	$i+j+k-il$	$-ij-il$	$-j+l$	$-1-k$	$i-ik$

Notes. $S = i + j + k + l$. **Every sacred vector is obtainable from one of the above by sign-changes, coordinate rotations, and postmultiplication by quaternion units.**

TABLE 2

The Middle Orbit Revealed. The vectors displayed are named (in order):

$$\begin{cases} \alpha, \alpha j, \alpha k, \alpha l \\ \beta, \beta j, \beta k, \beta l \\ \gamma, \gamma j, \gamma k, \gamma l \\ \delta, \delta j, \delta k, \delta l. \end{cases}$$

Every middle orbit vector can be obtained from one of the above by sign-changes, coordinate rotations, and multiplication by some power of i.

$1+j+k+l$	$-1+j+k-l$	$-1-j+k+l$	$-1+j-k+l$
0	0	0	0
0	0	0	0
$i+j+ik-l$	$-1+ij-k-il$	$-i+j+ik+l$	$1+ij-k+il$
0	0	0	0
$i+ij-k+l$	$-i+ij+k+l$	$1-j+ik+il$	$-1-j-ik+il$
$i-j+k+il$	$1+ij+ik-l$	$-1-ij+ik-l$	$-i+j+k+il$

$1+j-k-l$	$-1+j-k+l$	$1+j+k+l$	$1-j-k+l$
0	0	0	0
0	0	0	0
$i-j+ik+l$	$1+ij+k-il$	$-i-j+ik-l$	$-1+ij+k+il$
0	0	0	0
$i-ij+k+l$	$i+ij+k-l$	$-1-j+ik-il$	$-1+j+ik+il$
$i-j-k-il$	$1+ij-ik+l$	$1+ij+ik-l$	$i-j+k+il$

$1-j+k-l$	$1+j-k-l$	$-1+j+k-l$	$1+j+k+l$
0	0	0	0
0	0	0	0
$i+j-ik+l$	$-1+ij+k+il$	$i-j+ik+l$	$-1-ij-k+il$
0	0	0	0
$i-ij-k-l$	$i+ij-k+l$	$1+j+ik-il$	$1-j+ik+il$
$i+j-k+il$	$-1+ij+ik+l$	$1-ij+ik+l$	$-i-j-k+il$

$1-j-k+l$	$1+j+k+l$	$1-j+k-l$	$-1-j+k+l$
0	0	0	0
0	0	0	0
$i-j-ik-l$	$1+ij-k+il$	$i+j+ik-l$	$1-ij+k+il$
0	0	0	0
$i+ij+k-l$	$-i+ij-k-l$	$-1+j+ik+il$	$1+j-ik+il$
$i+j+k-il$	$-1+ij-ik-l$	$-1+ij+ik+l$	$i+j-k+il$

TABLE 3

Effect of Generators for $4R$ on the Typical Vector (q_n)

	I	J	K	L	M_0	N_0	P	Q	R	S
$q'_0 =$	$q_0 i$	$q_0 j$	$q_0 k$	$q_0 l$	$-q_0$	$q_0 i$	q_6	$q_0(jkl)$	$q_0(kl)$	$4q'_n = Sq_{-n}+$
$q'_1 =$	$q_1 i$	$q_1 j$	$q_1 k$	$q_1 l$	q_1	q_1^j	q_0	$q_4(jkl)$	$q_2(ilj)$	$+(i-l)q_{1-n}$
$q'_2 =$	$q_2 i$	$q_2 j$	$q_2 k$	$q_2 l$	q_2	q_2^k	q_1	$q_1(jkl)$	$q_1(ijl)$	$+(i-j)q_{2-n}$
$q'_3 =$	$q_3 i$	$q_3 j$	$q_3 k$	$q_3 l$	$-q_3$	$q_3 j$	q_2	$q_5(jkl)$	$q_6(ijlk)$	$+(l-k)q_{3-n}$
$q'_4 =$	$q_4 i$	$q_4 j$	$q_4 k$	$q_4 l$	q_4	q_4^l	q_3	$q_2(jkl)$	$q_4(ik)(jl)$	$+(i-k)q_{4-n}$
$q'_5 =$	$q_5 i$	$q_5 j$	$q_5 k$	$q_5 l$	$-q_5$	$q_5 l$	q_4	$q_6(jkl)$	$q_5(il)$	$+(k-j)q_{5-n}$
$q'_6 =$	$q_6 i$	$q_6 j$	$q_6 k$	$q_6 l$	$-q_6$	$q_6 k$	q_5	$q_3(jkl)$	$q_3(iklj)$	$+(j-l)q_{6-n}$

The further sign-changes M_n are defined by the condition that M_n negates the coordinates $q_n, q_{n+3}, q_{n+5}, q_{n+6}$ and fixes other q_m. The further operations N_n are transforms of N_0 by powers of P: N_n postmultiplies q_n by i, transforms q_{n+1} by j, and so on. We have $N_n^2 = M_n$, $[N_n, N_m] = \pm M_p$, for $p = p(m, n)$.

We have found it convenient in this table to write the typical quaternion not as $a + bj + ck + dl$, but rather as

$$q = ai + bj + ck + dl$$

Then (for instance) we define

$$q(ilj) = al + bi + ck + dj$$

the quaternion obtained from q by applying the permutation (ilj) to the units i, j, k, l. It is also convenient to define $q^j = j^{-1}qj$ etc., so that q^j is obtained from q by negating the k and l coordinates, and similarly q^k by negating the j and l coordinates, and q^l by negating the j and k ones.

We shall now describe these operations in more detail. The centre of $4R$ is the cyclic subgroup generated by i. Modulo this, the elements J, K, L define a 4-group. The element M_n negates $q_n, q_{n+3}, q_{n+5}, q_{n+6}$, and so the 7 elements of this form constitute the non-zero elements of an elementary Abelian group of order 8, which we call the sign-change group, whose elements commute with I, J, K, L.

Modulo the group so far, the elements N_n generate an elementary Abelian group of order 64. Here N_0 is displayed in the table, and N_n is N_0 transformed by that rotation of the coordinates that makes N_n multiply q_n by i. In fact we have the relations $N_n{}^2 = M_n$, and $[N_m, N_n] = \pm M_p$, for some $p = p(m, n)$. So the elements I, J, K, L, M, N generate a 2-group, of structure

$$\begin{matrix} 4\,. & 2^2 & .\,2^3 & .\,2^6 \\ I & J,K,L & M_n & N_n \end{matrix}$$

which we call the *diagonal group,* since it consists of those elements for which each q_n' depends only on the corresponding q_n.

The element P is our original element of order 7, and the element Q of order 3 normalises $\langle P \rangle$, while R of order 2 normalises $\langle Q \rangle$. These satisfy

$$P^7 = Q^3 = R^2 = 1, Q^{-1}PQ = P^2, R^{-1}QR = Q^{-1}, (PR)^3 = 1$$

which show that they generate the simple group $L_2(7)$ of order 168. This group normalises the diagonal subgroup, and extends it to the full monomial group, which therefore has the structure

$$\begin{matrix} 4\,. & 2^2 & .\,2^3 & .\,2^6 & .\,L_2(7) \\ I & J,K,L & M_n & N_n & P,Q,R \end{matrix}$$

and the order $2^{16}.3.7$. In the simple group R, this becomes $2^{14}.3.7$, and the structure simplifies to

$$\begin{matrix} 2^5 & .\,2^6 & .\,L_2(7) \\ J,K,L,M_n & N_n & P,Q,R \end{matrix}$$

(In these notes on structure, we use p^n for the elementary Abelian group of that order, n for a cyclic group of order n, and $A\,.\,B$ denotes a group with a normal subgroup A whose quotient is B. $[n]$ would be used for an *arbitrary* group of order n.)

Some more elements of the diagonal group are displayed in Table 4. Here N_{pqr} denotes an element that differs from $N_pN_qN_r$ only by sign-changes and scalar multiplications. Every diagonal element can be obtained from one in Table 4 by sign-changes (M_n), coordinate rotations, and postmultiplications by quaternion units.

5. The group 4R

The full normaliser of $\langle P \rangle$ is generated by its centraliser together with the elements Q and S, and we have the relations

$$Q^{-1}PQ = P^2, \quad Q^{-1}JQ = K, \quad Q^{-1}KQ = L, \quad Q^{-1}LQ = J, \quad Q^3 = 1$$

$$S^{-1}PS = P^{-1}, \quad JS = SJ, \quad KS = SK, \quad LS = SL, \quad QS = SQ, \quad S^2 = 1$$

which in particular show that Q and S generate a Frobenius group of order 42 with P. When we adjoin this new element S to the monomial group, we obtain the full group $4R$.

TABLE 4

The Diagonal Subgroup Revealed

N_0	N_{124}	N_{356}	N_{56}	N_{35}	N_{63}	N_{056}	N_{035}	N_{063}
$q'_0 = q_0 i$	q_0	$q_0 = q_0$	$q_0{}^{l}i$	$q_0{}^{j}i$	$q_0{}^{k}i$	$q_0{}^{l}i$	$q_0{}^{j}i$	$q_0{}^{k}i$
$q'_1 = q_1{}^{j}$	$q_1{}^{l}ik$	$q_1{}^{k}k = kq_1$	$q_1{}^{k}ij$	$q_1 ik$	$q_1{}^{k}il$	$q_1{}^{l}j$	$q_1{}^{j}k$	$q_1{}^{l}l$
$q'_2 = q_2{}^{k}$	$q_2{}^{j}il$	$q_2{}^{l}l = lq_2$	$q_2{}^{l}ij$	$q_2{}^{l}ik$	$q_2 il$	$q_2{}^{j}j$	$q_2{}^{j}k$	$q_2{}^{k}l$
$q'_3 = q_3 j$	$q_3{}^{l}k$	$q_3{}^{l}il = ilq_3$	$q_3{}^{l}il$	$q_3 l$	$q_3{}^{l}$	$q_3{}^{l}k$	$q_3 ik$	$q_3{}^{l}ij$
$q'_4 = q_4{}^{l}$	$q_4{}^{k}ij$	$q_4{}^{j}j = jq_4$	$q_4 ij$	$q_4{}^{j}ik$	$q_4{}^{j}il$	$q_4{}^{l}j$	$q_4{}^{k}k$	$q_4{}^{k}l$
$q'_5 = q_5 l$	$q_5{}^{k}j$	$q_5{}^{k}ik = ikq_5$	$q_5 k$	$q_5{}^{k}$	$q_5{}^{k}ik$	$q_5 ij$	$q_5{}^{k}il$	$q_5{}^{k}j$
$q'_6 = q_6 k$	$q_6{}^{j}l$	$q_6{}^{i}ij = ijq_6$	$q_6{}^{j}$	$q_6{}^{j}ij$	$q_6 j$	$q_6{}^{j}ik$	$q_6{}^{j}l$	$q_6 il$

The element N_{pqr} is congruent to $N_p N_q N_r$ modulo sign-changes and multiplication by units, and is the unique such element for which we need no minus signs. Every element of the diagonal group can be obtained from the identity or one of those listed by sign-changes, coordinate rotation, and multiplication by quaternion units.

Since S commutes elementwise with our quaternion group G, we have $(vq)S = (vS)q$ for any vector v and quaternion q, so that it must be possible to write S as a 7×7 matrix of quaternion left-multiplications. (Quaternion left-multiplications are the only 4-dimensional linear transformations that commute with the quaternion right-multiplications.) Moreover, since S anticommutes with P, this matrix must have a circulant property, and so is known as soon as we know $e_0 S$. This made the discovery of S fairly easy. Note that the coefficients for S appearing in Table 3 are the coordinates of one of the standard vectors. This vector, and its images under multiplications by P^n, J, K, and L, constitutes the unique other natural base permuted by these elements (to within scalar multiplications by powers of i).

6. Combinatorial properties of the sacred vectors

The vectors $4e_0$, $4e_0 j$, $4e_0 k$, $4e_0 l$ have the property that any other sacred vector which is orthogonal to three of them is orthogonal to the fourth. (This follows from the fact that any non-zero q_n coordinate of any long or middle orbit vector involves at least two of $1, j, k, l$.) We call such a set of four mutually orthogonal vectors a *quartet.*

Now it is not hard to show that the group contains elements sufficient to take any pair of orthogonal sacred vectors onto any other pair. We now ask how many quartets a given pair belong to? Taking the pair as $(4e_1, 4e_1 k)$ we find fairly easily that there are just five other pairs,

$$(4e_1 j, 4e_1 l), (4e_2 j, 4e_2 k), (4e_2, 4e_2 l), (4e_4, 4e_4 j), \text{ and } (4e_4 k, 4e_4 l)$$

which complete it to a quartet. Moreover, the union of any two of these six pairs is a quartet, so that the relation subsisting between the six pairs is symmetrical. We call the twelve vectors involved in such a system a *dozen.*

Now any quartet a, b, c, d belongs to three dozens, obtained from the three pairings $(a, b), (c, d)$ or $(a, c), (b, d)$ or $(a, d), (b, c)$. Each alternative gives us a set of 8 vectors, so that the original quartet determines 24 new vectors in this way, which complete it to a system of 28 mutually orthogonal vectors, which we call a *base.* To prove this, we need only consider the case of the quartet $(4e_0, 4e_0 j, 4e_0 k, 4e_0 l)$. One of the three completions of this to a dozen is by the adjunction of the eight similar vectors with the subscript 0 replaced by 1 or 3, another by those with 0 replaced by 2 or 6, and the third by those with 0 replaced by 4 or 5, and so this quartet has led to our original base.

It turns out that there are just three distinct kinds of triple of three mutually orthogonal sacred vectors:

(i) *quartic triples,* consisting of three vectors from a quartet,
(ii) *twelvic triples,* three vectors which belong to a dozen but no quartet,
(iii) *basic triples,* which belong to a base but no dozen.

In each case, the relevant quartet, dozen, or base is uniquely determined by the three vectors. The subgroup of $4R$ which fixes each of the three 1-spaces as a whole has order $2^{13}, 2^{10}, 2^7$ in the three respective cases (corresponding to subgroups of order $2^{11}, 2^8$, and 2^5 in R).

With these exceptions, the group $4R$ is transitive on triples of vectors with given scalar products. The subgroup of $4R$ which fixes the three relevant 1-spaces as a whole has order

(iv) 800 if each pair from the triple has negative scalar product,
(v) 192 if each pair has positive scalar product,
(vi) 208 if each pair has complex scalar product,
(vii) 64 if the triple has just one pair of orthogonal vectors,
(viii) 80 if the triple has just two orthogonal pairs.

The cases (i)–(viii) cover all the orbits of $4R$ on triples of vectors, to within the possibilities of replacing any vector by a scalar multiple of itself.

In case (iv), the three vectors are part of a set of 5 which have zero sum. There is an interesting configuration of 20 vectors which contains many such fives and which can be determined in various ways from a few of its members. One such *twenty* is made from the four quartets

$$(4e_0, 4e_0 j, 4e_0 k, 4e_0 l), \quad (\alpha, \alpha', \alpha'', \alpha''')$$
$$(\beta l, \beta l', \beta l'', \beta l'''), \quad (\gamma j, \gamma j', \ldots), \quad (\delta k, \delta k', \ldots)$$

where the primes indicate vectors obtained from the given ones by all possible sign-changes, and $\alpha, \beta, \gamma, \delta$ are the vectors displayed in Table 2.

The most interesting of the various groups that arise, apart from the monomial group attached to a given base, is the *dozen group.* We note that there are seven dozens in a base, which convert the seven quartets that determine that base into a seven point projective plane – see Fig. 1, where E_n denotes the base $(4e_n, 4e_n j, 4e_n k, 4e_n l)$ and the lines denote dozens.

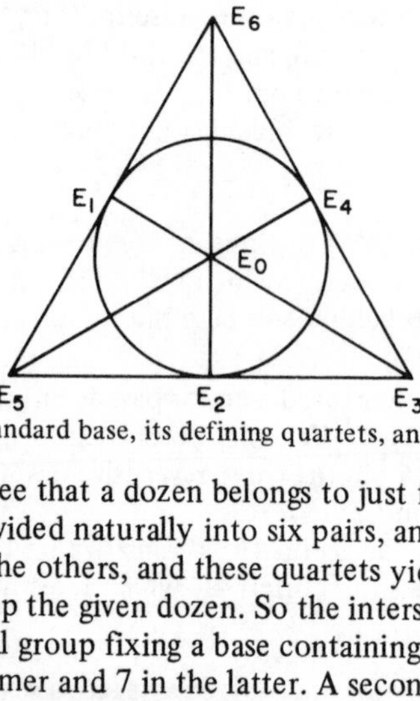

Fig. 1 The standard base, its defining quartets, and its dozens.

But it is also easy to see that a dozen belongs to just five bases, for the twelve vectors in a dozen are divided naturally into six pairs, any one of which extends to five quartets by adding the others, and these quartets yield bases, three of whose defining quartets make up the given dozen. So the intersection of the dozen group with the monomial group fixing a base containing the dozen is a group having index 5 in the former and 7 in the latter. A second base containing the dozen $E_1 \cup E_2 \cup E_4$ is shown in Fig. 2.

In fact this dozen group is the centraliser of an involution, for the sign-change M_0 fixes the 12-space defined by the dozen $E_1 \cup E_2 \cup E_4$ and negates the complementary 16-space. Abstractly the dozen group has shape

$$\underset{I}{4.2} \;.\; \underset{M_0}{} \; \underset{J, K, L, N_n}{[2^{10}]} \; .PGL_2(5)$$

The normal 2-subgroup consists of the operations fixing each of the 12 1-spaces as a whole, and the quotient group is realised as a group of permutations on the 6 pairs comprising the dozen. (At the moment of writing, I am unable to prove

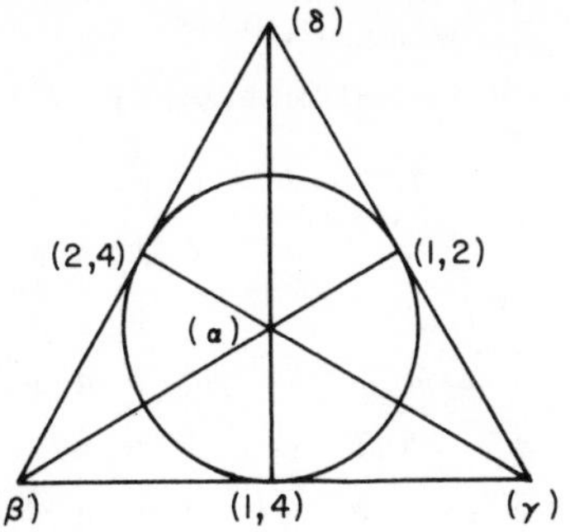

Fig. 2 A base meeting the standard base in the (1, 2, 4) dozen – (γ) denotes the quartet of four vectors obtained from the vector γ of Table 2 by sign-changes. The three remaining quartets in the figure are:

$(1, 2) = \{4e_1, 4e_1k, 4e_2j, 4e_2k\}$
$(2, 4) = \{4e_1, 4e_2l, 4e_4k, 4e_4l\}$
$(1, 4) = \{4e_4, 4e_4j, 4e_1j, 4e_1l\}$

this assertion about the structure of the dozen group, for which there is an alternative possibility of a normal 2-group of twice the given order, with quotient $PSL_2(5)$. But I believe the asserted possibility to be the actual one.)

The fact that the dozen-group is the centraliser of an involution leads us to an interesting and amusing way of working inside $4R$. Any two orthogonal vectors define a dozen, and so define an involution fixing the space defined by that dozen and negating the orthogonal complement (this is the *negative* of the involution we earlier considered). We shall call this involution *the reflection in the dozen,* or, since it is determined by the given pair of orthogonal vectors, *the reflection in that pair.* So $4R$ is rather like one of Fischer's transposition groups, except that with Fischer's groups the objects naturally permuted are themselves involutions, whereas in $4R$ we need *two* of the permuted vectors to determine an involution.

Now whenever we have a configuration of vectors, we can consider the further vectors obtained by transforming some vector of the configuration by the reflection in some orthogonal pair of the configuration. A number of interesting configurations and subgroups of $4R$ can be found in this way.

7. The subgroup $U_3(5)$

One of the vectors of Table 1 is

$$(S,\ 1-il,\ 1-ij,\ -ik-il,\ 1-ik,\ -ij-ik,\ -ij-il)$$

which by coordinate rotations yields a set of seven sacred vectors any two of which have scalar product 4. In fact $4R$ contains symmetries realising all the

TABLE 5

50 Vectors Permuted by $U_3(5)$

	i	j	k	l	i	j	k	l	i	j	k	l	i	j	k	l
q_0:	b	b	b	b	c	c	b	c	c	b	c	c	c	c	c	b
q_1:	b	b	b	b	c	b	b	a	b	a	a	c	c	a	a	b
q_2:	b	b	b	b	b	c	a	a	c	b	a	a	a	a	b	b
q_3:	b	b	b	b	a	a	b	c	b	b	a	a	b	b	a	a
q_4:	b	b	b	b	c	a	b	a	a	b	a	b	b	a	c	a
q_5:	b	b	b	b	b	a	a	b	a	b	c	a	b	a	a	b
q_6:	b	b	b	b	b	a	b	a	b	a	b	a	a	c	a	b

	i	j	k	l	i	j	k	l	i	j	k	l	i	j	k	l
q_0:	d	d	d	d	b	b	b	b	b	b	b	b	b	c	c	c
q_1:	a	b	b	a	a	a	d	a	a	d	a	a	a	b	b	a
q_2:	a	a	b	b	a	a	a	d	a	a	d	a	a	a	b	b
q_3:	b	b	a	a	a	d	a	a	d	a	a	a	a	a	c	b
q_4:	a	b	a	b	a	d	a	a	a	a	a	d	a	b	a	b
q_5:	b	a	a	b	a	a	a	d	d	a	a	a	a	c	b	a
q_6:	b	a	b	a	a	a	d	a	d	a	a	a	a	b	a	c

The coordinates of these vectors are given as linear combinations $Ai + Bj + Ck + Dl$ (*not* $A + Bj + Ck + Dl$). Only the numbers A, B, C, D are displayed. We use the abbreviations $a = 1 + 2i$, $b = 1 - 3i$, $c = -4 + 2i$, $d = -4 - 3i$. Two of these vectors are *adjacent* when their difference is 5 times a sacred vector, and each of the 50 is adjacent in this sense to just 7 others, so that just 175 sacred vectors (to within negation) arise in this way.

To obtain the complete set of 50 vectors from those listed, cyclically permute the coordinates.

even permutations of these seven vectors, and so contains a subgroup A_7 which preserves their sum. This sum is the vector in which each quaternion coordinate is $(1 - 3i)(i + j + k + l) = (1 - 3i)S$ – let us call it V.

The vector V determines the seven sacred vectors of which it is the sum. If v is any one of them, then $V - 5v$ is another vector like V, which is the sum of seven sacred vectors in just one way ($-v$ being one of them), the seven vectors being permuted by a group A_7 fixing $V - 5v$. We say that V is *adjacent* to $V - 5v$. Repeatedly taking adjacent vectors in this way, it can be verified that we eventually get a set of 50 vectors like V, each adjacent to just seven others, the adjacency graph being the so-called Hoffman-Singleton graph [3]. The difference of any two adjacent vectors from this set of 50 is 5 times a sacred vector, and we obtain, to within scalar multiplication by -1, just 175 sacred vectors in this way.

We have said enough to prove that the Rudvalis group contains $U_3(5)$, which is a subgroup of index 2 in the full automorphism group of the Hoffman-Singleton graph. In fact it is not hard to write down the 50 images of V in full, since they fall into just 8 orbits under P. For the benefit of readers who might wish to study the embedding of $U_3(5)$ in R in more detail, we have therefore listed these 50 vectors in Table 5. The existence of the subgroup $U_3(5)$ in R has also been demonstrated independently by K-C. Young [4], who has also found a subgroup isomorphic to $PSL_2(29)$ in R.

Perhaps we should remark that $4R$ also contains $P \Sigma U_3(5)$, in which $U_3(5)$ appears to index 2, and which extends the above group A_7 to a group S_7 permuting the same 7 vectors as the A_7. This S_7 is obtained from A_7 by adjoining our element S.

References

1. J. H. Conway and D. B. Wales. *J. Algebra* **3**(1973), 538–548.
2. A. Rudvalis. Private communication.
3. A. J. Hoffman and R. R. Singleton. *I.B.M. Jour. Res. Dev.* **4**(1960), 497–504.
4. K-C.Young. *Notices Amer. Math. Soc.*, **155**(August 1974), pA-481.
5. U. Dempwolff. *J. Algebra* **1**(1974), 53–88.

(Reference 5 is not mentioned in the text, but has much information about R, and several other references.)

5. Finite Permutation Groups, Edge-coloured Graphs and Matrices

PETER M. NEUMANN

The Queen's College, Oxford

This paper is an expansion of five expository lectures on finite permutation groups. My aim was to explain some recent combinatorial ideas due mainly to Charles C. Sims and D. G. Higman, and some related matrix theory, the so-called centraliser-ring theory, whose origins seem to lie further back, in the work of Issai Schur and of J. Sutherland Frame. I have restricted myself to the elementary parts of the theory and very little of what I have included is really new. Nevertheless, I hope these notes will be useful in introducing future enthusiasts to an enjoyable part of algebra, and also as a commentary on Chapters III and V of the great little monograph [48] by Helmut Wielandt.

1. Permutation groups

Throughout these lectures Ω is to denote a finite set with n elements, and G a group of permutations of Ω, that is, $G \leqslant \mathrm{Sym}(\Omega)$. The relation on Ω defined by specifying that $\alpha \sim \beta$ if there exists g in G such that $\alpha g = \beta$ is an equivalence relation. Its equivalence classes are known as the *orbits* of G in Ω; the orbit containing α is denoted by α^G, and it consists of all the images of α under permutations in G: that is, $\alpha^G = \{\alpha g | g \in G\}$. If $\Omega_1, \ldots, \Omega_k$ are the orbits then G induces a permutation group G^{Ω_i} on Ω_i, and, since the construction of G as permutation group from its constituents G^{Ω_i} is relatively easy ($\Omega = \Omega_1 \cup \ldots \cup \Omega_k$, and G is a sub-direct product of the groups $G^{\Omega_1}, \ldots, G^{\Omega_k}$), for most purposes it is enough to have information about these constituents. Therefore we specialise immediately and assume that G has just one orbit in Ω: thus for all $\alpha, \beta \in \Omega$ there exists g in G so that $\alpha g = \beta$; and G is said to be *transitive* on Ω.

If, given $\alpha_1, \alpha_2, \beta_1, \beta_2$ in Ω, with $\alpha_1 \neq \alpha_2$ and $\beta_1 \neq \beta_2$, there always exists g in G such that $\alpha_1 g = \beta_1$ and $\alpha_2 g = \beta_2$, then G is said to be *doubly transitive* (or 2-transitive, or 2-fold transitive: we define k-fold transitivity analogously). Following Burnside I shall say that G is *simply transitive* if G is transitive but not doubly transitive. The subject of these lectures concerns techniques appropriate specifically for the study of simply transitive groups.

For a point α of Ω the *stabiliser* G_α is defined by $G_\alpha := \{g \in G \mid \alpha g = \alpha\}$. This is a subgroup of G and, given that G is transitive, there is a natural one-one correspondence between Ω and the set of its cosets: if $\beta \in \Omega$ then $\{g \in G \mid \alpha g = \beta\}$ is a coset $G_\alpha h$ (where h is any element of G such that $\alpha h = \beta$). Consequently $|G : G_\alpha|$ $= |\Omega| = n$, where $|G : G_\alpha|$ is the index of G_α in G. Other stabilisers are all conjugate in G to G_α: if $\alpha h = \beta$, then $G_\beta = h^{-1} G_\alpha h$. The importance of the stabilisers arises from the fact that the action of G on Ω is isomorphic in a natural sense to the action of G on the set of cosets of G_α by right multiplication. Therefore the embedding of G_α in G gives information about the action of G on Ω. If $G_\alpha = 1$ then G is said to be *regular* on Ω. In this case Ω may be identified with G and the action is simply that of right multiplication, the "Cayley representation" of G on itself.

An invariant equivalence relation, or *congruence*, is an equivalence relation ρ on Ω such that if $\alpha \equiv \beta \pmod{\rho}$ then $\alpha g \equiv \beta g \pmod{\rho}$ for all g in G. The "universal" relation, in which $\alpha \equiv \beta$ for all α, β in Ω, and the "trivial" relation, of equality, are congruences. If there are no other congruences then G is said to be *primitive* on Ω. (We are assuming that G is transitive. In any case, transitivity follows from the given condition for primitivity unless $|\Omega| = 2$ and G is the trivial group.) If G is imprimitive and $\Theta_1, \ldots, \Theta_m$ are the equivalence classes for a non-trivial proper congruence ρ on Ω, then G permutes the sets Θ_i bodily. Since G is transitive on Ω it also acts transitively on the set $\Omega/\rho := \{\Theta_1, \ldots, \Theta_m\}$, so all the sets Θ_i are of the same size, l say, and $n = lm$. The sets Θ_i are known as *blocks* of imprimitivity. The study of G can usually be reduced to the study of its two constituents, the group of degree m induced by G on Ω/ρ, and the group of degree l induced by the set-wise stabiliser of one of the blocks upon it. In terms of the stabilisers it is not hard to see that (given transitivity of G on Ω) G is primitive if and only if the stabiliser G_α is a maximal proper subgroup of G. In fact, there is an isomorphism between the lattice of congruences on Ω and the lattice of those subgroups of G which contain G_α. Notice that if G is 2-transitive then it certainly is primitive: some of the deeper theorems of permutation group theory give the converse under suitable conditions.

One obtains a considerable amount of information by studying the action of G on ordered pairs and the action of a stabiliser G_α on Ω. The two are closely related as the following lemma shows.

Lemma 1. *Given α in Ω (and given that G is transitive on Ω) there is a one-one correspondence between orbits $\Delta_0, \Delta_1, \ldots, \Delta_{r-1}$ of G acting on $\Omega \times \Omega$ and orbits $\Gamma_0, \Gamma_1, \ldots, \Gamma_{r-1}$ of the stabiliser G_α in Ω. The numeration can be so chosen that*

$$\Gamma_i = \Gamma_i(\alpha) := \{\gamma \mid (\alpha, \gamma) \in \Delta_i\}.$$

Proof. What we need to show is that the sets $\Gamma_i(\alpha)$ defined in the statement are the orbits of G_α. Notice that, because of the transitivity of G on Ω, Δ_i does contain pairs of the form (α, γ), and so $\Gamma_i(\alpha) \neq \emptyset$. If $\gamma_1, \gamma_2 \in \Omega$ then (α_i, γ_1) and (α, γ_2) are equivalent under G if and only if there exists g in G such that $\alpha g = \alpha$ and $\gamma_1 g = \gamma_2$. Thus (α, γ_1) and (α, γ_2) lie in the same orbit of G acting on $\Omega \times \Omega$ if and only if γ_1 and γ_2 lie in the same orbit of G_α in Ω, as required.

The orbits $\Gamma_0, \Gamma_1, \ldots, \Gamma_{r-1}$ of G_α are known as the *suborbits* of G. Some authors prefer the notion of *orbital* (see, for example, D. G. Higman [16]): an orbital is a function Γ defined on Ω and with values in the set of subsets of Ω, having the two properties that $\Gamma(\alpha)$ is an orbit of G_α in Ω and that if $\alpha g = \beta$ then $\Gamma(\alpha)g = \Gamma(\beta)$. Orbitals are consistent identifications of the suborbits considered as orbits of different stabilisers and so they are the functions Γ_i described in the last line of the lemma. The common number r of suborbits and of G-orbits in $\Omega \times \Omega$ is known as the *rank* of G (D. G. Higman [15]). The rank is 2 if and only if G is doubly transitive. We assign the indices so that Δ_0 is the diagonal $\{(\omega, \omega) | \omega \in \Omega\}$, and then Γ_0 is $\{\alpha\}$, the *trivial* suborbit.

For any subset Δ of $\Omega \times \Omega$ we put $\Delta^* := \{(\beta, \alpha) | (\alpha, \beta) \in \Delta\}$. Then Δ^* is a G-orbit if and only if Δ is a G-orbit. Thus we can define an involution (permutation of order 2) on $\{0, 1, \ldots, r-1\}$ such that $\Delta_{i^*} = \Delta_i^*$. The corresponding suborbits Γ_i and Γ_{i^*} (denoted also by Γ_i^*) are said to be *paired*. This pairing is the same as that described by H. Wielandt in ([48], Section 16).

The sizes $n_i := |\Gamma_i|$, often called the "lengths" of the suborbits, are known as the *subdegrees* of G. We have that $n_0 = 1$ and $n_{i^*} = n_i$, and we usually order the suborbits so that $n_0 \leqslant n_1 \leqslant \ldots \leqslant n_{r-1}$.

2. Edge-coloured graphs and their automorphism groups

When I talk of a *graph* I have in mind a set V of "vertices" and a set E of "edges", where $E \subseteq V \times V$. Thus my graphs (usually) are directed graphs without multiple edges: if α, β are vertices then there is at most one edge "from" α "to" β. Moreover, I shall assume that graphs have no loops, that is, that $E \subseteq V \times V - \Delta_0$, where $\Delta_0 := \{(\omega, \omega) | \omega \in V\}$. We will sometimes use a name for the vertex set also as a name for the graph.

The automorphism group Aut (V) of the graph is $\{g \in \mathrm{sym}\,(V) | Eg = E\}$. It is said to be transitive, or "vertex transitive" if it is transitive on V. It is already quite a strong restriction on the graph V that it admits a transitive group of automorphisms. For example, we have

Lemma 2. *If the graph V is finite and admits a transitive automorphism group then every component of V is strongly connected.*

Before proving it I should explain what this lemma means. If $\alpha, \beta \in V$ then an *undirected path* from α to β is a sequence $\alpha_0, \alpha_1, \ldots, \alpha_l$ of vertices such that, for all relevant i, either (α_{i-1}, α_i) or (α_i, α_{i-1}) is an edge, and in which $\alpha_0 = \alpha$, $\alpha_l = \beta$. For a *directed path* we require that (α_{i-1}, α_i) is an edge for $1 \leqslant i \leqslant l$. The *length* of the path is the number l. The relation ρ such that $\alpha \equiv \beta \pmod \rho$ if and only if there is an undirected path from α to β is an equivalence relation (in fact, it is the equivalence relation generated by E), and its equivalence classes are the *components* of V. The subset W of V is said to be *strongly connected* if, for all α, β in W, there is a directed path from α to β. Here is a portrait of an example whose components are not strongly connected:

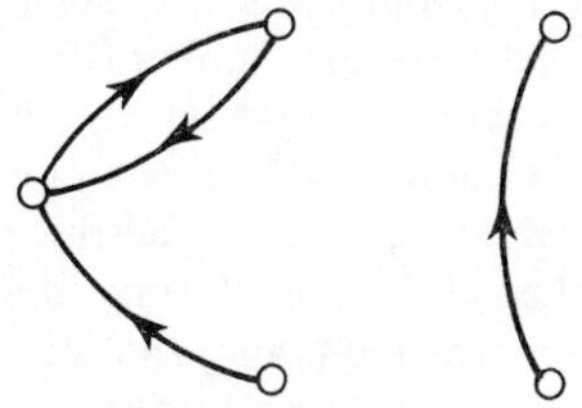

Proof of Lemma 2. It will be sufficient to show that, if there is an edge from vertex α to vertex β, then there is some directed path from β to α. Put

$$T(\alpha) := \{\omega \in V \mid \text{there is a directed path from } \alpha \text{ to } \omega\}.$$

If $\beta \in T(\alpha)$ and $\gamma \in T(\beta)$ then there are directed paths from α to β and from β to γ, and their concatenation is a directed path from α to γ: thus $T(\beta) \subseteq T(\alpha)$. Since V admits a transitive group of automorphisms there is some automorphism g such that $\alpha g = \beta$; then $T(\alpha)g = T(\beta)$ and consequently $T(\beta)$ has the same number of elements as $T(\alpha)$; therefore, from the inclusion proved in the preceding sentence, together with the finiteness of V, it follows that $T(\beta) = T(\alpha)$. This has shown that if there is a directed path from α to β then, since $\alpha \in T(\alpha)$, there is a directed path from β to α, which is quite enough to prove the lemma.

The *complete* graph of order n is that graph V with n vertices in which $E = V \times V - \Delta_0$. We are interested in complete graphs in which a partition of E as the disjoint union of subsets $E_1, \ldots, E_s$ is prescribed. I think of this as an *edge-coloured* complete graph: select s colours $c_1, \ldots, c_s$ and assign to edge (α, β) the colour c_i if $(\alpha, \beta) \in E_i$. The graphs V with edge sets E_i will be referred to as the *monochrome* subgraphs of this edge-coloured complete graph. The automorphism group of the coloured graph is the group of all permutations g of V such that (α, β) and $(\alpha g, \beta g)$ have the same colour for all α, β in V.

Return now to the situation where G is a transitive group of permutations of the set Ω. The orbits $\Delta_1, \ldots, \Delta_{r-1}$ of G acting on $\Omega \times \Omega - \Delta_0$ provide a partition of the edge-set of the complete graph on Ω and so we have an edge-coloured complete graph associated with G. To each non-trivial suborbit Γ_i of G there corresponds a colour c_i. The corresponding monochrome subgraph is the "orbital graph" introduced by C. C. Sims in [38]. The coloured complete graph associated with G offers facilities for studying not only each individual suborbit through its orbital graph, but also the way in which these orbital graphs fit together to form the complete directed graph.

An edge-coloured complete graph which arises in this way from a group has several special properties. For example, its automorphism group (which is what Wielandt in [49] calls the "2-closure" of G) is transitive on vertices and on the edges of each monochrome subgraph; the stabiliser of a vertex is transitive on the edges of each colour emanating from it; each monochrome subgraph is regular, in the sense that every vertex has the same number (the valency) of edges of that colour emanating from it (or terminating in it), and in fact the valency is the appropriate subdegree; furthermore, if (Ω, Δ) is a monochrome subgraph then so is (Ω, Δ^*), the graph obtained by reversing all the edges in Δ. If all the suborbits of G are self-paired then each monochrome subgraph is completely determined by the underlying non-directed graph, and we consider the corresponding edge-coloured complete graph in the ordinary (undirected) sense.

Examples

A_3 of degree 3:

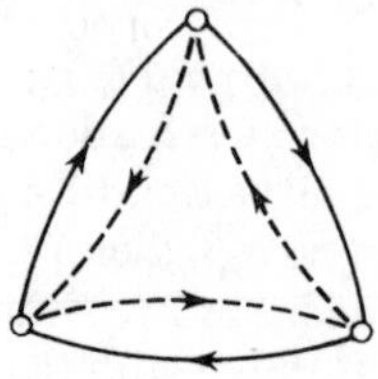

D_8 of degree 4:
(all suborbits self-paired)

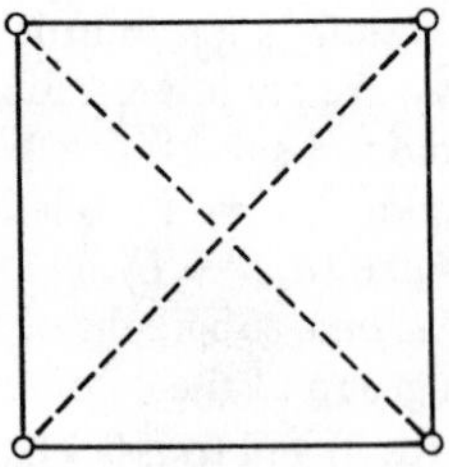

S_5 of degree 10 acting on unordered pairs from $\{1, 2, 3, 4, 5\}$. This group is of rank 3 and its suborbits are all self-paired:

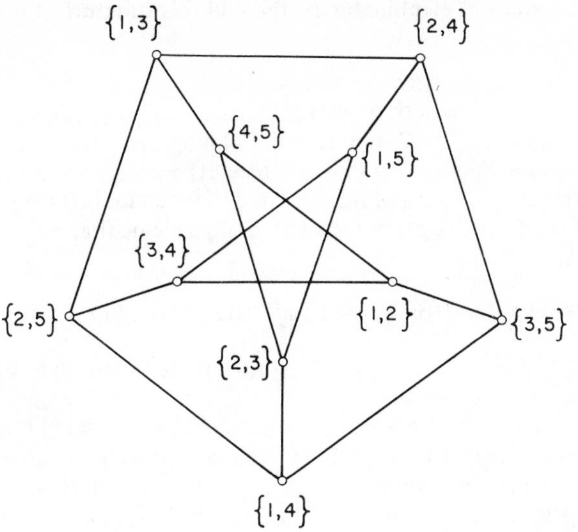

I have drawn the monochrome subgraph corresponding to one of the non-trivial suborbits. It is the famous Petersen graph. The other is of course simply the complementary graph.

To finish this lecture here is a splendidly useful criterion for primitivity of G due to D. G. Higman (see [38], Prop. 4.4, or [16], 1.12):

Lemma 3. *G is primitive on Ω if and only if every monochrome subgraph is connected.*

Proof. By Lemma 2, each component of each monochrome subgraph is strongly connected, and so the statement is unambiguous. The relation of being joined by a path of colour c_i is a G-congruence on Ω and so if G is primitive then every monochrome subgraph must be connected. Conversely, suppose that G is imprimitive and that Θ is a block of imprimitivity. Since $|\Theta| > 1$ there exist distinct points α, β in Θ. Suppose that (α, β) has colour c_i. If $\gamma \in \Theta$ and (γ, δ) is an edge of colour c_i then there exists g in G with $\alpha g = \gamma$ and $\beta g = \delta$; then $\gamma \in \Theta \cap \Theta g$, so $\Theta = \Theta g$; hence $\beta g \in \Theta$, that is, $\delta \in \Theta$. This shows that every edge of colour c_i emanating from a point of Θ terminates in Θ and consequently Θ is a union of components of the c_i-monochrome subgraph. Since $\Theta \neq \Omega$ it follows that this monochrome subgraph is not connected, which completes the proof.

3. Some applications of combinatorics: the suborbit structure of primitive groups†

We suppose that G is transitive on Ω, that $\alpha \in \Omega$, that G_α has orbits $\Gamma_0, \Gamma_1, \ldots, \Gamma_{r-1}$ and that G has corresponding orbits $\Delta_0, \Delta_1, \ldots, \Delta_{r-1}$ in $\Omega \times \Omega$, as in Section 1. Notice that if $r = 2$, that is, if G is doubly transitive, then the coloured graph introduced in Section 2 is monochrome, its automorphism group is the full symmetric group on Ω, and from its study we cannot learn anything about the action of G on Ω. Therefore I shall generally assume that $r \geqslant 3$. For subsets Θ, Φ of $\Omega \times \Omega$ we define

$$\Theta \circ \Phi := \{(\alpha, \beta) | \alpha \neq \beta \text{ and there exists } \gamma \text{ in } \Omega \text{ such that } (\alpha, \gamma) \in \Theta \text{ and } (\gamma, \beta) \in \Phi\}.$$

Thus the subgraph with edge-set $\Delta_i \circ \Delta_j$ contains the edge (α, β) if and only if there is a path of length 2 in our coloured graph Ω, directed from α to β, whose first step has colour c_i and whose second step is coloured c_j. It is one of the very special properties of the coloured graph arising from G, not shared by coloured graphs in general, that $\Delta_i \circ \Delta_j$ is either empty (which happens only if $j = i^*$ and the subdegree (valency) n_i is 1) or it is a union of monochrome subgraphs. We shall write $\Gamma_i \circ \Gamma_j$ for the appropriate union of suborbits of G.

Almost all of the applications of these combinatorial ideas refer to primitive groups. Here is the first of them.

Theorem 1 (C. Jordan, 1870). *If G is primitive on Ω and Γ is a non-trivial suborbit, then every composition factor of G_α is isomorphic to a section (a factor group of a subgroup) of G_α^Γ.*

The point is that the action of G_α on Γ might well have a non-trivial kernel K, and on the face of it the suborbit Γ can only tell us about the group G_α^Γ, that is, G_α/K. But the theorem guarantees that the composition factors of K are not too wild.

Proof. I shall use Γ also to denote the orbital graph, that is, the monochrome subgraph of Ω corresponding to the given suborbit. Since G is primitive, by Lemmas 2 and 3 this graph is strongly connected and we can define "distance" in Ω by

$$d(\omega_1, \omega_2) := \min \{l | \text{there is a directed } \Gamma\text{-path of length } l \text{ from } \omega_1 \text{ to } \omega_2\}$$

† For a more advanced survey see Cameron [6].

(but beware: it is not always true that $d(\omega_1, \omega_2) = d(\omega_2, \omega_1)$ unless Γ is self-paired). We define the "diameter" d of our graph by

$$\begin{aligned} d &:= \max \{ d(\omega_1, \omega_2) | \omega_1, \omega_2 \in \Omega \} \\ &= \max \{ d(\alpha, \omega) | \omega \in \Omega \} \end{aligned}$$

and we define "discs" and "circles" about α by

$$\Theta_a := \{ \omega | d(\alpha, \omega) \leqslant a \}, \qquad \Sigma_a := \{ \omega | d(\alpha, \omega) = a \}.$$

Now let

$$K_a := G_{(\Theta_a)} := \bigcap_{\omega \in \Theta_a} G_\omega.$$

Then certainly we have

$$G_\alpha = K_0 \geqslant K_1 \geqslant K_2 \geqslant \ldots \geqslant K_d = 1.$$

Moreover, since Θ_a is a union of G_α-orbits and K_a is the kernel of the action of G_α on Θ_a, we have $K_a \lhd G_\alpha$ for all a. Now $K_{a-1}/K_a \cong K_{a-1}^{\Sigma_a}$. Let Φ be a K_{a-1}-orbit

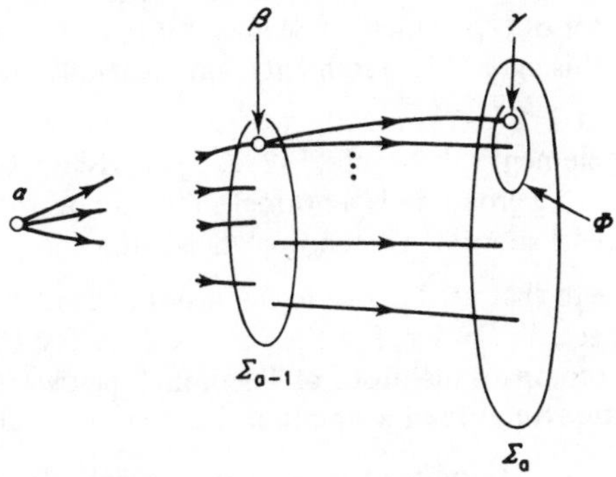

in the circle Σ_a, and let $\gamma \in \Phi$. Then there exists β in Θ_{a-1} (in fact, $\beta \in \Sigma_{a-1}$) such that $\gamma \in \Gamma(\beta)$. Then, since $K_{a-1} \leqslant G_\beta$ we have that $\Phi \subseteq \Gamma(\beta)$, and K_{a-1}^{Φ} is a homomorphic image of $K_{a-1}^{\Gamma(\beta)}$, which is a subgroup of $G_\beta^{\Gamma(\beta)}$. But of course $G_\beta^{\Gamma(\beta)} \cong G_\alpha^{\Gamma}$. Since $K_{a-1}^{\Sigma_a}$ is a subdirect product of the groups K_{a-1}^{Φ} where Φ ranges over K_{a-1} orbits in Σ_a, we have that K_{a-1}/K_a is a sub-direct product of groups isomorphic to sections of G_α^{Γ}. Therefore, since every composition factor of G_α is a composition factor of K_{a-1}/K_a for some a (by the Jordan-Hölder Theorem), we have the result.

Corollaries: *Suppose that G is primitive and* Γ *is a non-trivial suborbit.*

(1) *If* G_α^Γ *is a* π*-group for some set* π *of primes then* G_α *is a* π*-group.*

(2) *If* G_α^Γ *is soluble then so is* G_α.

(3) *If the suborbits are numbered so that* $n_0 \leqslant n_1 \leqslant n_2 \leqslant \ldots \leqslant n_{r-1}$ *then, if p is a prime divisor of* n_i *(some* $i \geqslant 1$*), then* $p \leqslant n_1$.

Notice that the proof shows a little more than these. For example we can sharpen (2) to

($2^\#$) *If* G_α^Γ *is soluble then so is* G_α *and* $l(G_\alpha) \leqslant d.l(G_\alpha^\Gamma)$ *where l denotes derived length and d is the diameter of the* Γ*-graph.*

The original source of the theorem appears to be [20], Théorème 395, pp. 281–284. Jordan's statement is different from ours since he wrote in terms of polynomials, and since by "facteur de composition" he meant merely the order of what we now call "composition factor" ([20], p. 42). However, the translation from polynomials to groups is automatic, and indeed Jordan's proof is written in terms of the appropriate Galois groups. In [24], Section 37, p. 83 W. A. Manning proves Corollary 1 by a different method, and his idea can quite easily be used to prove the whole theorem; a short proof can also be found in [48], Theorem 18.2.

One is tempted to hope that every composition factor of G_α appears actually as a composition factor of G_α^Γ (which de Séguier [36], p. 85, proves and attributes to Jordan), but this agreeable statement is not generally true:

Exercise. Suppose that $3 \leqslant k < \frac{1}{2}m$, and let G be the symmetric group S_m acting on the set Ω of all k-element subsets of $\{1, 2, \ldots, m\}$. Show that G is primitive on Ω, that the alternating group A_k is a normal subgroup of a stabiliser G_α, but that there is a non-trivial suborbit Γ such that A_k is *not* a composition factor of G_α^Γ.

Arguments similar to that used above in the proof of Jordan's Theorem, sometimes using the sets $\Gamma, \Gamma \circ \Gamma^*, \Gamma \circ \Gamma^* \circ \Gamma, \Gamma \circ \Gamma^* \circ \Gamma \circ \Gamma^*, \ldots$ in place of the discs and circles (compare the proof of Theorem 2 below), give further detail on the group-theoretic structure of a stabiliser. Here are two such points:

Exercise. Show that if G is primitive and has a non-trivial suborbit Γ of prime length p then p divides $|G_\alpha|$ and p^2 does not (Rietz [31]).

Exercise. Show that if G is primitive and has a non-trivial suborbit Γ such that both G_α^Γ and $G_\alpha^{\Gamma^*}$ are regular then G_α is faithful on Γ and on Γ^* (i.e. $|G_\alpha| = |\Gamma|$; this particular result can also be proved easily by a direct group-theoretic argument).

The next theorem deals with the order of magnitude of the suborbits in a primitive group. It is stated in a slightly stronger form than is usual ([48], Theorem 17.4, or [38], Prop. 4.5).

Theorem 2. *Suppose that G is primitive on Ω and that the suborbits are numbered so that $n_0 \leqslant n_1 \leqslant n_2 \leqslant \ldots \leqslant n_{r-1}$. If $n_1 = 1$ then $n_i = 1$ for all i (and so G is regular on Ω). If $n_1 > 1$ then $n_i \leqslant (n_1 - 1) n_{i-1}$ for $i \geqslant 2$.*

Proof. If $n_1 = 1$ then it is almost immediate that each component of the Γ_1-graph (the c_1-monochrome graph on Ω) must be a directed "cycle". By Lemmas 2 and 3 this graph must be strongly connected and so it is just one cycle of length n. Then, an automorphism fixing the point α fixes every point of the cycle, so $G_\alpha = 1$, G is regular on Ω, and $n_i = 1$ for all i. (Furthermore n must be prime and G cyclic of order n: see Theorem 5, below.)

Now suppose that $n_1 > 1$, and assume also that $n_i > (n_1 - 1)n_{i-1}$ for some $i \geqslant 2$. Put $\Sigma := \Gamma_0 \cup \Gamma_1 \cup \ldots \cup \Gamma_{i-1}$. Define a "zig-zag c_1-path" to be a directed path whose edges are coloured alternately c_1 and c_1*, starting with c_1:

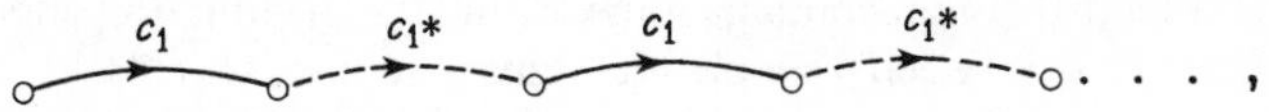

that is,

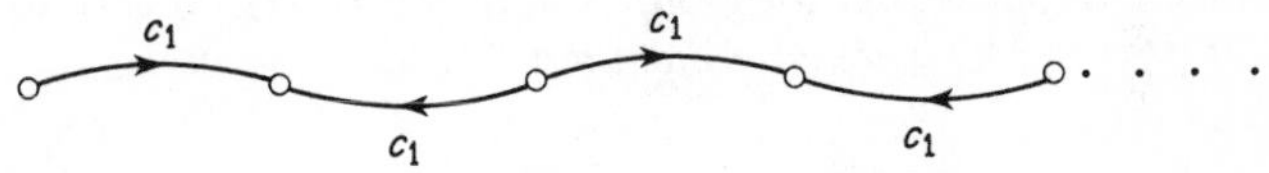

We define

$$\Sigma_m := \{\omega \mid \text{there is a zig-zag } c_1\text{-path of length at most } m \text{ from } \alpha \text{ to } \omega\},$$

and we shall use induction to prove that $\Sigma_m \subseteq \Sigma$ for all m. Certainly this is true if m is 0 or 1, so, as inductive hypothesis we assume that it is true for all $m \leqslant l$, where $l \geqslant 1$. Suppose that $\gamma \in \Sigma_{l+1}$. We may take it that the shortest zig-zag c_1-path from α to γ is of length $l + 1$: so if (β, γ) is the last step in such a path then $\beta \in \Sigma_l$, and (β, γ) has colour c_1 if l is even, colour c_1* if l is odd. Accordingly, if j, k are the indices such that $\beta \in \Gamma_j$ and $\gamma \notin \Gamma_k$, then $\gamma \in \Gamma_j \circ \Gamma_1$

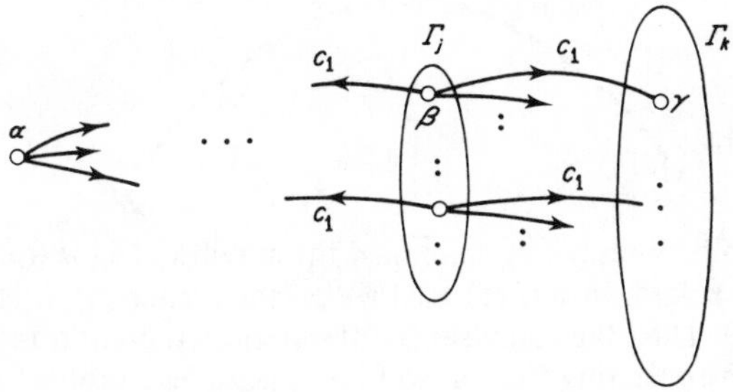

(or $\gamma \in \Gamma_j \circ \Gamma_{1^*}$), and so $\Gamma_k \subseteq \Gamma_j \circ \Gamma_1$ (or $\Gamma_k \subseteq \Gamma_j \circ \Gamma_{1^*}$). Now there are just $n_1 n_j$ edges of colour c_1 (or c_{1^*}) emanating from points of Γ_j; but from the definition of Σ_l in terms of zig-zag c_1-paths it should be clear that at least one such edge from each point of Γ_j terminates in a point of Σ_{l-1}; therefore at most $(n_1 - 1)n_j$ of these edges can terminate in Γ_k; consequently $n_k \leqslant (n_1 - 1)n_j$. Now $\beta \in \Sigma_l$, so, by inductive assumption, $\beta \in \Sigma$, and this means that $j \leqslant i-1$. Therefore $n_k \leqslant (n_1 - 1)\, n_{i-1} < n_i$, hence $k \leqslant i-1$ and so $\gamma \in \Sigma$. This shows that $\Sigma_{l+1} \subseteq \Sigma$ and, by induction, that $\Sigma_m \subseteq \Sigma$ for all m.

Finally, since $n_1 > 1$ we know that $\Gamma_1 \circ \Gamma_{1^*}$ is a non-empty union of suborbits; however, what we have just shown is more than enough to prove that the component containing α of the graph corresponding to $\Gamma_1 \circ \Gamma_{1^*}$ is contained in Σ. Thus there are monochrome subgraphs which are not connected and this contradicts Lemma 3.

Our next theorem contains some arithmetic information about the subdegrees. First we need a simple preliminary result:

Lemma 4. *Suppose that the group X acts transitively on each of the sets Ω_1 and Ω_2. If $|\Omega_1|$ and $|\Omega_2|$ are co-prime then X is transitive acting on $\Omega_1 \times \Omega_2$ in the natural way.*

Proof. Let Δ be an orbit of X in $\Omega_1 \times \Omega_2$. Since X is transitive on Ω_1 the size of the set $\Delta(\alpha)$, where $\Delta(\alpha) := \{\beta | (\alpha, \beta) \in \Delta\}$, does not depend on the element α of Ω_1. Therefore $|\Omega_1|$ divides $|\Delta|$. Similarly of course $|\Omega_2|$ divides $|\Delta|$. Consequently $|\Omega_1| \,.\, |\Omega_2|$ divides $|\Delta|$ and so $\Delta = \Omega_1 \times \Omega_2$.

Theorem 3 (Marie J. Weiss, 1935). *Suppose that the subdegrees n_i, n_j of G are co-prime, and that $n_i > n_j$. Then $\Gamma_i \circ \Gamma_j$ is a single suborbit Γ_k; moreover n_k divides $n_i n_j$ and $n_k \geqslant n_i$; if G is primitive and $n_j > 1$, then $n_k > n_i$.*

Proof. Consider the action of G on the set Θ of triples $(\alpha', \gamma', \beta')$ where (α', γ') has colour c_i and (γ', β') is coloured c_j. Since the orbits $\Gamma_i^*(\gamma')$ and $\Gamma_j(\gamma')$ have

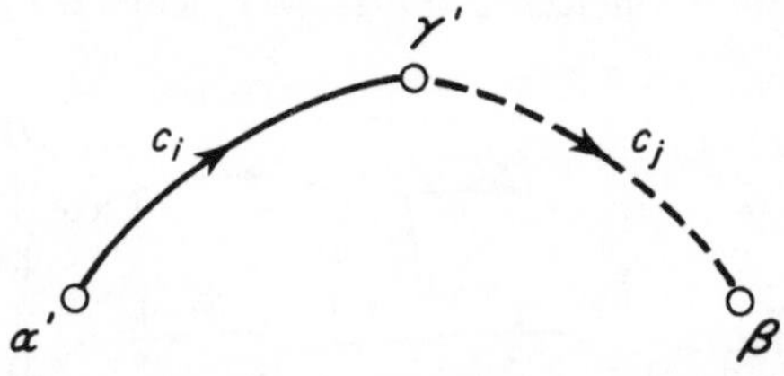

sizes n_i, n_j, which are co-prime, by Lemma 4 the stabiliser $G_{\gamma'}$ is transitive on pairs such as (α', β'). And since G is transitive on the vertices γ', it follows that G is transitive on Θ. Then the stabiliser G_α of our chosen point α is transitive on configurations in Θ emanating from α, so $\Gamma_i \circ \Gamma_j$ is a single orbit of G_α.

Next, count triangles (α, γ, β) in Θ.

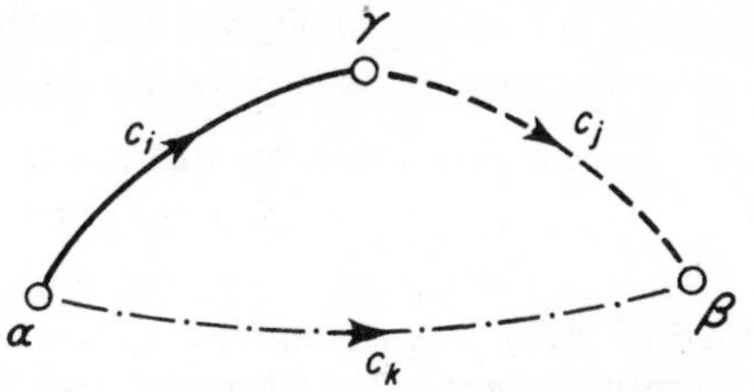

Counting in two different ways (keeping α fixed, allowing β, γ to vary) shows that $n_i n_j = n_k . x$ where x is the number of such triangles on a fixed base (α, β) of colour c_k. Thus n_k divides $n_i n_j$, and since obviously $x \leqslant \min(n_i, n_j) = n_j$, we have $n_k \geqslant n_i$.

Suppose finally that $n_k = n_i$. Then $x = n_j$, and this means that $\Gamma_j^*(\beta) \subseteq \Gamma_i(\alpha)$ for all β in $\Gamma_k(\alpha)$. Therefore, if $\beta \in \Gamma_k(\alpha)$ then $\Gamma_j^* \circ \Gamma_j(\beta) \subseteq \Gamma_k(\alpha)$. If $n_j > 1$ then $\Gamma_j^* \circ \Gamma_j$ is a non-empty union of suborbits and the foregoing sentence tells us that $\Gamma_k(\alpha)$ is a union of components of its graph. Thus, by Lemma 3, if G is primitive and $n_j > 1$ then $n_k > n_i$.

Notice that if G is primitive and $n_j = 1$ then either $j = 0$ (in which case trivially $k = i$), or G is regular and of prime degree (see Theorem 2, or Theorem 5, below). As a consequence of the theorem we have the

Corollaries: (1) *In a primitive group which is not regular of prime degree the largest of the subdegrees* n_{r-1} *has a divisor in common with each of the other non-trivial subdegrees.*
(2) *If a primitive group has k pairwise co-prime non-trivial subdegrees then its rank is at least* 2^k.
(3) *A primitive group of degree* $p + 1$, *where p is prime and* $p \geqslant 3$, *cannot have rank 3.*

Groups of small rank with non-trivial co-prime subdegrees appear to be rather rare. It would be interesting to know whether the lower bound for the rank in Corollary 2 can ever be attained if $k \geqslant 2$, and in particular we have

Problem 1 (P. J. Cameron). *Does there exist a primitive group of rank 4 with two of its non-trivial subdegrees co-prime?*

The small Janko group of order 175,560 has a primitive representation of degree 266 and rank 5, with subdegrees 1, 11, 12, 110, 132 (see [23], or [5]).

Since the basis of Theorem 3 was Lemma 4 it can be generalised considerably, and such generalisations are often useful. The four conditions

(i) X is transitive on $\Omega_1 \times \Omega_2$;
(ii) X is transitive on Ω_1 and, if $\alpha \in \Omega_1$, then X_α is transitive on Ω_2;

(iii) X is transitive on Ω_2 and, if $\beta \in \Omega_2$, then X_β is transitive on Ω_1;
(iv) X is transitive on Ω_1 and Ω_2, and if $\alpha \in \Omega_1, \beta \in \Omega_2$, then $X_\alpha X_\beta = X_\beta X_\alpha = X$

are equivalent, and I shall say that Ω_1, Ω_2 are *co-prime X-spaces* to describe this situation. Now Theorem 3 can be extended as follows:

Suppose that the suborbits Γ_i, Γ_j of G are co-prime G_α-spaces, and that $n_i \geqslant n_j$. Then $\Gamma_i^* \circ \Gamma_j$ is a single suborbit Γ_k; moreover n_k divides $n_i n_j$ and $n_k \geqslant n_i$; if G is primitive and $n_j > 1$, then $n_k > n_i$.

(N.B. In Theorem 3 we used $\Gamma_i \circ \Gamma_j$ since n_i and n_{i^*} are the same, but in general, when Γ_i and Γ_j are co-prime G_α-spaces it need not be true that Γ_i^* and Γ_j are co-prime.)

Lemma 4 gave a purely arithmetic condition for Ω_1 and Ω_2 to be co-prime X-spaces. A rather different condition, which is often useful in this context of the suborbits of primitive groups, is: if X is doubly transitive on Ω_1 and transitive on Ω_2, if either $|\Omega_2| < |\Omega_1|$ or $|\Omega_2| = |\Omega_1|$ and X is simply transitive on Ω_2, then Ω_1 and Ω_2 are co-prime X-spaces. This follows easily from some elementary character theory: if X is transitive on Ω_1 and on Ω_2, then Ω_1 and Ω_2 are co-prime X-spaces if and only if the "permutation characters" π_1 and π_2 of X which Ω_1 and Ω_2 afford have only the principal character in common as an irreducible constituent. Using this criterion we get as a significant improvement of Corollary 1 that in a non-regular primitive group the permutation character of G_α afforded by the longest suborbit Γ_{r-1} must have a non-principal irreducible constituent in common with the characters afforded by each of the other non-trivial suborbits.

Theorem 4 (W. A. Manning, P. J. Cameron). *Suppose that G is simply primitive*† *on Ω and that Γ is a non-trivial suborbit with subdegree v, on which G_α is doubly transitive. Then $\Gamma^* \circ \Gamma$ is a single self-paired suborbit and it has subdegree w, where w divides $v(v-1)$ and $w \geqslant 2(v-1)$.*

Proof. As in the proof of Theorem 3 we observe that G is transitive on the set Θ of all triples $(\alpha', \gamma', \beta')$ with $\gamma' \in \Gamma^*(\alpha')$ and $\beta' \in \Gamma(\gamma')$, for, G is transitive on the

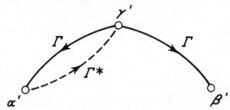

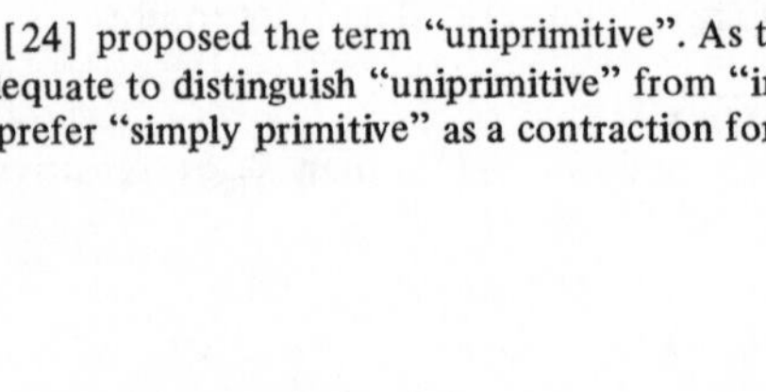

† W. A. Manning in [24] proposed the term "uniprimitive". As there are few scholars with calligraphic skills adequate to distinguish "uniprimitive" from "imprimitive", and as I am not one of those few, I prefer "simply primitive" as a contraction for "simply transitive and primitive".

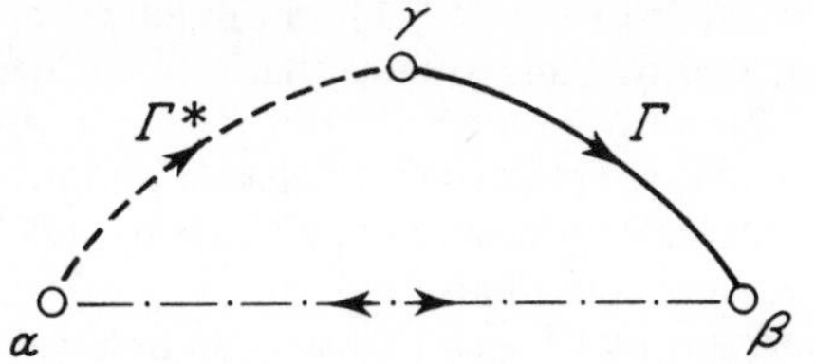

vertices γ' and, by assumption, the stabiliser $G_{\gamma'}$ is transitive on pairs such as (α', β') in $\Gamma(\gamma')$. Now if we fix α then we see that G_α is transitive on $\Gamma^* \circ \Gamma$. Moreover it is clear from its definition that $\Gamma^* \circ \Gamma$ is self-paired. Counting triangles (α, γ, β) in Θ gives that $v(v-1) = x.w$ where x is the number of such triangles on a given base (α, β) in $\Gamma^* \circ \Gamma$. Thus w divides $v(v-1)$. Now suppose that $x > \frac{1}{2}v$. Then, if $\beta_1, \beta_2 \in (\Gamma^* \circ \Gamma)(\alpha)$, then $\Gamma^*(\beta_1) \cap \Gamma^*(\alpha)$ and $\Gamma^*(\beta_2) \cap \Gamma^*(\alpha)$ both contain more

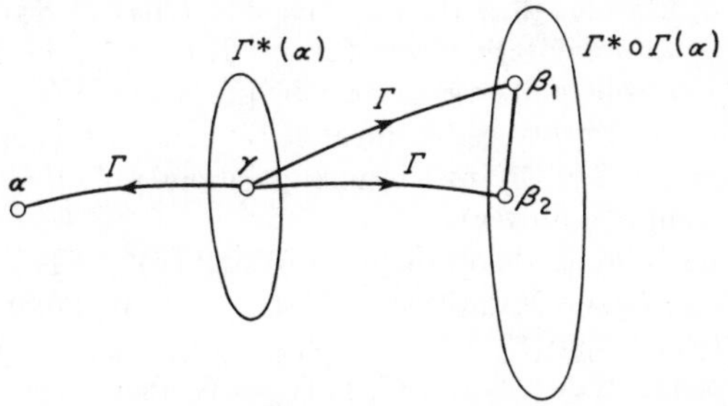

than a half of the points in $\Gamma^*(\alpha)$ and so they must intersect; there then exists γ in $\Gamma^*(\alpha)$ such that $(\gamma, \beta_1) \in \Gamma$ and $(\gamma, \beta_2) \in \Gamma$; consequently $\beta_2 \in (\Gamma^* \circ \Gamma)(\beta_1)$; therefore every edge between points of $\Gamma^* \circ \Gamma$ is in the $(\Gamma^* \circ \Gamma)$-graph and $\{\alpha\} \cup (\Gamma^* \circ \Gamma)$ is a component of this graph. By Lemma 3 that is not possible and so $x \leqslant \frac{1}{2}v$. Therefore $w \geqslant 2(v-1)$, and the proof is complete.

The origins of this theorem can be traced back to a paper of Elizabeth R. Bennett [1], Corollary II of Theorem V, and to the lectures of W. A. Manning [24], Section 38, but it was not until 1929 that Manning showed (in [26], where he also traces the history further back, to a rather obscure passage in a paper of Camille Jordan (1873)) how to prove that, if there is a doubly transitive suborbit of length v, with $v > 2$, then there is another suborbit, whose length w exceeds v and divides $v(v-1)$. The proof given above, due independently to P. J. Cameron [3] and to W. L. Quirin [30], is very much simpler than Manning's and lends itself nicely to further extension: in [4], II, Cameron has proved an inequality of the form $w > O(v^{6/5})$, while in [4], I, he has shown that if G_α is 3-fold transitive on Γ then in general (that is, with some interestingly special exceptions) w must be $v(v-1)$ or $\frac{1}{2}v(v-1)$. In this latter paper he also uses a lovely argument originated by Sims to show that *if G_α is 2-transitive on the suborbit Γ then it is also 2-transitive on* Γ^*. A stronger result of this nature is to be found in Knapp's paper [21]. Further articles on primitive groups with multiply transitive suborbits have been written by Cameron, Eiichi Bannai, Hikoe Enomoto, Wolfgang Knapp, and others.

The last of the theorems that I wish to discuss in this lecture concern short suborbits. Here is an easy introduction:

Theorem 5. *Suppose that G is primitive on Ω.*
(i) *If G has a non-trivial suborbit of length 1 then G is regular and cyclic of prime order and degree.*
(ii) *If G has a suborbit of length 2 then G is dihedral of order $2n$ and n is prime.*

Proof. We have already (Theorem 2) seen a combinatorial proof that if there is a non-trivial suborbit of length 1 then G is regular. This can also be understood easily in purely group-theoretic terms, for, if $\{\beta\}$ is a G_α-orbit then $G_\beta = G_\alpha$, hence $g^{-1}G_\alpha g = G_\alpha$ where $\beta = \alpha g$; so that the normaliser $N(G_\alpha)$ contains G_α properly and, by maximality of G_α, must be G; thus $G_\alpha \lhd G$ and consequently $G_\alpha = 1$. Now the trivial subgroup 1 is maximal if and only if G is a cyclic group of prime order, and this proves (i).

There is a similar group-theoretic proof of part (ii): if β lies in a G_α-orbit of length 2 then $|G_\alpha : G_{\alpha\beta}| = 2$; similarly, $|G_\beta : G_{\alpha\beta}| = 2$; therefore $G_{\alpha\beta} \lhd G_\alpha$ and $G_{\alpha\beta} \lhd G_\beta$, so $N(G_{\alpha\beta}) \geqslant gp(G_\alpha, G_\beta)$; since G_α, G_β are maximal proper subgroups of G and are unequal they generate G, so $G_{\alpha\beta} \lhd G$; then, being contained in a stabiliser, $G_{\alpha\beta}$ must be trivial; now G is generated by two cyclic subgroups G_α, G_β of order 2 and is therefore dihedral of order $2n$; the maximality of G_α finally implies that n is prime. An alternative proof that $G_{\alpha\beta} = 1$, that is, that G_α is of order 2 comes from the exercises on page 90, but it is perhaps worth seeing how one may derive the fact that G is dihedral by a direct combinatorial argument. First observe that if Γ is a suborbit of length 2 then $\Gamma^* \circ \Gamma$ is also a suborbit of length 2 and is self-paired. (Actually one can show that n must be odd and then that all the suborbits must be self-paired.) Then G is a group of automorphisms of an undirected graph of valency 2 which is connected (by D. G. Higman's Theorem, Lemma 3). Such a graph is simply a polygon and its automorphism group is easily seen to be dihedral of order $2n$.

On page 51 of his book [48] Wielandt asks what can be concluded from the existence of a suborbit of length 3. It was mainly in order to answer this question that Sims introduced these graph-theoretic ideas in [38]. He observed that if the supposed suborbit is self-paired then the corresponding orbital graph is "cubic" (that is, regular of valency 3) and that W. T. Tutte [44, 45] had already produced interesting results on the automorphism groups of cubic graphs. By adding to Tutte's methods a clever calculation with 2-groups Sims was able to handle directed graphs of valency 3 and to prove

Theorem 5(iii) (C. C. Sims, 1967). *If G is primitive and has a suborbit of length 3 then* $|G_\alpha|$ *divides* 3.2^4.

Starting from this result and using some of the deep classifications of finite simple groups W. J. Wong has calculated a complete list [51] of the groups *G* which can arise, and so Wielandt's question has been satisfactorily answered.

Theorem 5 supports the following

Conjecture (C. C. Sims). *There is a function f on the natural numbers such that if G is primitive and has a suborbit of length d then* $|G_\alpha| \leqslant f(d)$.

The original ideas of Tutte and Sims have been further developed by A. J. Gardiner, W. Knapp, W. L. Quirin, Sims himself and R. M. Weiss, particularly for the cases where d is 4 or 5. J. G. Thompson [43] has proved in a quite different way that there is a prime p such that at least the index of $O_p(G_\alpha)$ (the largest normal p-subgroup of G_α) in G_α is bounded by a function of d. In [50] Wielandt showed how his theory of subnormal subgroups could be applied to give an elegant proof of Thompson's theorem and some more besides. Very recently W. Knapp [22], using a combination of the combinatorial methods of Sims, the group-theoretic ideas of Wielandt, and some new ingredients of his own, has finally settled the conjecture for the cases $d = 4$, $d = 5$ and for certain other cases where d is prime:

Theorem 5 (W. Knapp, 1976). *Suppose that G is primitive and has a suborbit of length d.*

(iv) *If d is 4, then* $|G_\alpha|$ *divides* 2^5 *or* $2^4 . 3^6$;

(v) *If d is 5, then* $|G_\alpha|$ *divides* $5.3^2 . 2^{14}$;

(vi) *If d is a prime such that every transitive permutation group of degree d is either soluble or 2-primitive, then* $|G_\alpha|$ *divides* $d . (d-1)!^2$.

(To say that a group is 2-primitive means that it is 2-transitive and the stabiliser of one point is primitive on the remainder. By a theorem of W. Burnside every transitive group of prime degree is either soluble or 2-transitive, so the hypothesis of part (iii) is not as strong as it looks. It is satisfied by the primes 5, 11, 17, 19, 23, 29, 37, ...)

The theorems presented in this lecture have applications to several problems about permutation groups. One of their uses is in cataloguing the primitive groups of small degree.

Exercise. Describe all the simply primitive groups of degree n with $n \leqslant 19$.

As a hint I shall finish the lecture with a computation to show that there are no simply primitive groups of degree 20. Of course such a result still leaves one with the task of finding all the multiply transitive groups: but that is another problem for which other techniques are appropriate and available.

Example. All primitive groups of degree 20 are doubly transitive.

Proof. Suppose on the contrary that G is a simply primitive group of degree 20. Since 19 is prime the rank of G is at least four (Theorem 3) and since 20 is not prime there cannot be non-trivial suborbits of length 1 or 2 (Theorem 5). A glance at Wong's list shows that there cannot be any suborbits of length 3 either (this can also be shown by more elementary, but more lengthy arguments based on Theorems 1 and 4; or by observing that, from a theorem of Rietz (see p. 90 above), the order of G would have to be $5.3.2^a$ for some a, that therefore G would have a transitive permutation representation of degree 15 on the cosets of a Sylow 2-subgroup, and that the structure of G follows easily). If the shortest non-trivial suborbit has length 4 then G_α cannot act on it as a 2-group, otherwise, by Theorem 1, G_α would be a 2-group, all the non-trivial suborbits would have even length, and 2 would divide 19, which is not the case; therefore G_α is 2-transitive on this suborbit of length 4 and, by Theorem 4, there is also a suborbit of length 6 or 12; since there must be a non-trivial suborbit of odd length, and since the prime divisors of $|G_\alpha|$ are less than 4 (Theorem 1), the only possibility for the subdegrees is 1, 4, 6, 9; but this contradicts Theorem 3. If the shortest non-trivial suborbit had length 5 then, by Theorem 3, the longest would have length 5, 10 or 15 and it is easy to see that none of these are possible. Finally, if the shortest non-trivial suborbit has length 6 or more then, since the rank is at least four the subdegrees would have to be 1, 6, 6, 7, and this contradicts Theorem 3.

A convenient list of the primitive groups of degree n with $n \leqslant 20$ has been published by Sims ([40], see also [1] and references quoted there). The computation has been continued for $n \leqslant 50$ by Sims and his associates, but apart from announcements in [41, 30] the results have not yet been published.

4. Adjacency matrices and the centraliser ring

As in Section 1 we have G acting transitively on the set Ω, the orbits of a stabiliser G_α in Ω are $\Gamma_0, \Gamma_1, \ldots, \Gamma_{r-1}$, and the corresponding orbits of G in $\Omega \times \Omega$ are $\Delta_0, \Delta_1, \ldots, \Delta_{r-1}$. We define the corresponding *basic adjacency matrices* $B_0, B_1, \ldots, B_{r-1}$ to be n by n matrices with rows and columns indexed by Ω, where

$$(B_i)_{\alpha,\beta} := \begin{cases} 1 & \text{if } (\alpha, \beta) \in \Delta_i \\ 0 & \text{if } (\alpha, \beta) \notin \Delta_i. \end{cases}$$

Thus B_i is, in the usual sense of graph theory, the adjacency matrix of the monochrome subgraph (Ω, Δ_i). Notice that $B_0 = I$, and that B_{i^*} is the transpose B_i^{Tr}, therefore that B_i is symmetric precisely when the corresponding suborbit is self-paired. The fact that the monochrome graphs neatly cover the complete graph on Ω without overlapping is equivalent to the equation

$$B_0 + B_1 + \ldots + B_{r-1} = J$$

where J is the matrix all of whose coordinates are 1. Products of the adjacency matrices enumerate paths in Ω: if $M := B_{i_1} B_{i_2} \ldots B_{i_l}$ (and if $i_\nu > 0$ for all ν) then $M = (m_{\alpha,\beta})$ where $m_{\alpha,\beta}$ is the number of directed paths of length l from α to β whose first step has colour i_1, second step has colour $i_2, \ldots$, last step has colour i_l.

We identify G with a group of permutation matrices X_g in the usual way: $X_g := (g_{\alpha,\beta})$ where

$$g_{\alpha,\beta} := \begin{cases} 1 & \text{if } \alpha g = \beta \\ 0 & \text{if } \alpha g \neq \beta \end{cases}$$

and then we have the easy but significant

Lemma 5. *The set $\{B_0, B_1, \ldots, B_{r-1}\}$ is a basis for the space V of all matrices commuting with every element of G (the "centraliser ring" of G).*

Proof. If $M = (m_{\alpha,\beta})$ and $g \in G$ then

$$MX_g = \left(\sum_\nu m_{\alpha,\nu}\, g_{\nu,\beta} \right) = (m_{\alpha,\beta g^{-1}})$$

$$X_g M = \left(\sum_\nu g_{\alpha,\nu}\, m_{\nu,\beta} \right) = (m_{\alpha g,\beta}).$$

Thus M commutes with all the matrices X_g if and only if $m_{\alpha,\beta g^{-1}} = m_{\alpha g,\beta}$ for all $\alpha, \beta \in \Omega$ and all $g \in G$. Replacing β by βg shows that $m_{\alpha,\beta}$ must be constant over orbits of G in $\Omega \times \Omega$, and hence that M must be a linear combination of the basis matrices $B_0, B_1, \ldots, B_{r-1}$.

Corollary. *The space V spanned by the basic adjacency matrices $B_0, B_1, \ldots, B_{r-1}$ is an algebra. That is, there exist numbers a_{ijk} such that*

$$B_i B_j = \sum_{k=0}^{r-1} a_{ijk} B_k.$$

Up to now I have not specified where my matrices live. The lemma and its corollary are true over any commutative ring with unity. I shall assume for the remainder of these lectures that we work over a (cyclotomic number) field K of characteristic 0, which is a splitting field for G. In this case we know from general representation theory that V must be a semisimple algebra and that K is a splitting field for V. In fact, if the irreducible constituents of the matrix representation of G are

$$F_0, F_1, \ldots, F_{s-1}$$

of degrees

$$f_0, f_1, \ldots, f_{s-1}$$

appearing with multiplicities

$$e_0, e_1, \ldots, e_{s-1}\ ,$$

then (see, for example, [8], pp. 42, 49) there exists a unitary matrix U with coefficients in K such that

$$U^{-1}X_gU = \begin{pmatrix} F_0(g) & & & \\ & e_1\left\{\begin{matrix} F_1(g) & & \\ & \ddots & \\ & & F_1(g) \end{matrix}\right. & & \\ & & \ddots & \\ & & & e_{s-1}\left\{\begin{matrix} F_{s-1}(g) & & \\ & \ddots & \\ & & F_{s-1}(g) \end{matrix}\right. \end{pmatrix}$$

where the diagonal entries denote the f_i by f_i square matrices $F_i(g)$ and all other entries are 0. (N.B. We can choose the numbering so that F_0 is the principal representation: then $F_0(g) = 1$ for all g and $e_0 = f_0 = 1$.) The transform $U^{-1}VU$ consists of all the matrices commuting with the block diagonal matrices drawn above and so it is not hard to show that if $Y \in V$ then

$$U^{-1}YU = \begin{pmatrix} E_0(Y) & & & & \\ & E_1(Y) \otimes I_{f_1} & & & \\ & & \cdot & & \\ & & & \cdot & \\ & & & & \cdot \\ & & & & & E_{s-1}(Y) \otimes I_{f_{s-1}} \end{pmatrix}$$

where $E_0, E_1, \ldots, E_{s-1}$ are certain irreducible representations of V of degrees 1, $e_1, \ldots, e_{s-1}$, and where $\otimes$ denotes Kronecker product of matrices (thus the multiplicities of the constituents $E_0, E_1, \ldots, E_{s-1}$ in the natural matrix representation of V are $1, f_1, \ldots, f_{s-1}$).

Rather than considering the matrix representations themselves we usually work with their characters. Thus if π is the permutation character of G associated with its action on Ω, that is $\pi(g) := \operatorname{tr}(X_g) = |\operatorname{fix}_\Omega(g)|$, then

$$\pi = \chi_0 + \sum_1^{s-1} e_\lambda \chi_\lambda$$

where χ_0 is the principal character and χ_λ is the character of the matrix representation F_λ. Similarly, if η denotes the "natural" character of V, that is $\eta(Y) := \operatorname{tr}(Y)$ for $Y \in V$, then we have that

$$\eta = \eta_0 + \sum_1^{s-1} f_\lambda \eta_\lambda, \text{ where } \eta_\lambda(Y) := \operatorname{tr}(E_\lambda(Y)).$$

Collecting these facts together we have:

Lemma 6. *There is a one-one correspondence between irreducible constituents χ_λ of the permutation character π of G and irreducible constituents η_λ of the natural character η of V. If $\pi = \chi_0 + \sum_1^{s-1} e_\lambda \chi_\lambda$ where χ_λ is an irreducible character of degree f_λ, then $\eta = \eta_0 + \sum_1^{s-1} f_\lambda \eta_\lambda$ where η_λ is a character of degree e_λ. Moreover $\eta_0(Y)$ is the row sum (or column sum) of Y, and in particular, $\eta_0(B_i) = n_i$.*

The fact that the principal character χ_0 appears in π with multiplicity 1, and the fact that η_0 is as described in the last sentence of the lemma, can be seen from an easy calculation showing that the fixed point set of G acting on the vector space K^n is one-dimensional, spanned by $\mathbf{t} := (1, 1, \ldots, 1)$. It is equivalent to the well-known theorem of Frobenius that

$$\frac{1}{|G|} \sum_{g \in G} |\mathrm{fix}_\Omega(g)| = 1.$$

Notice that

$$1 + \sum_{1}^{s-1} e_\lambda f_\lambda = n$$

because this is the dimension of our matrix representation of G; and also

$$1 + \sum_{1}^{s-1} e_\lambda{}^2 = r$$

because this is the dimension of V.

The simplest examples of the reduction described on page 101 occur when all the multiplicities e_λ are 1. Then $s = r$, and our matrix representation of G is said to be *multiplicity-free.* In this case $U^{-1}YU$ is a diagonal matrix for all Y in V, the representations $E_\lambda(Y)$ and their characters $\eta_\lambda(Y)$ are 1-dimensional, that is, they are simply eigenvalues of Y appearing with multiplicities f_λ.

Incidentally, these facts show further how very special the coloured graph associated with G is. For, it will not generally be true that the basic adjacency matrices of an edge-coloured complete graph span an algebra, and if by chance they do, one would not expect this algebra to be semi-simple nor to have a splitting field whose Galois group is abelian.

Let us return now to simpler considerations about the matrices B_i and the multiplication constants a_{ijk}. Since B_i and B_j are non-negative integer matrices so is their product. If (α, β) is an edge of colour c_k then a_{ijk} must be the (α, β)-coordinate of B_iB_j. Hence the multiplication constants a_{ijk} are non-negative integers. In fact they have a clear combinatorial significance: Since $B_0 = I$ we have

$$a_{iok} = \begin{cases} 1 & \text{if } i = k \\ 0 & \text{if } i \neq k \end{cases}$$

$$a_{ojk} = \begin{cases} 1 & \text{if } j = k \\ 0 & \text{if } j \neq k, \end{cases}$$

and, since a_{ij0} is the number of points β with (α, β) coloured c_i and (β, α) coloured c_j,

$$a_{ij0} = \begin{cases} n_i & \text{if } j = i^* \\ 0 & \text{if } j \neq i^*. \end{cases}$$

Also, if none of i, j, k is 0, then a_{ijk} is the number of triangles (α, γ, β) on a fixed base (α, β) of colour c_k, whose other edges (α, γ), (γ, β) are coloured c_i, c_j respectively. Notice that therefore $n_k\, a_{ijk}$ is the number of oriented triangles at each vertex having edges of colours c_i, c_j, c_{k^*} in that order, and $nn_k a_{ijk}$ is the total number of such triangles in the coloured graph. Counting these and related triangles in different ways produces relations such as

$$\begin{aligned} n_k\, a_{ijk} &= n_i\, a_{jk^*i^*} = n_j\, a_{k^*ij^*} \\ &= n_k\, a_{j^*i^*k^*} = n_i\, a_{kj^*i} = n_j\, a_{i^*kj}. \end{aligned}$$

The less obvious relations

$$\sum_\nu a_{ij\nu}\, a_{\nu kl} = \sum_\mu a_{jk\mu} a_{i\mu l},$$

which are equivalent to the associativity of V, also have combinatorial significance. The left side enumerates quadrangles $(\alpha, \gamma, \delta, \beta)$ on a fixed edge (α, β) of colour c_l, whose other sides, (α, γ), (γ, δ), (δ, β) are coloured c_i, c_j, c_k respectively, by counting those in which the diagonal (α, δ) is coloured c_ν and summing. The

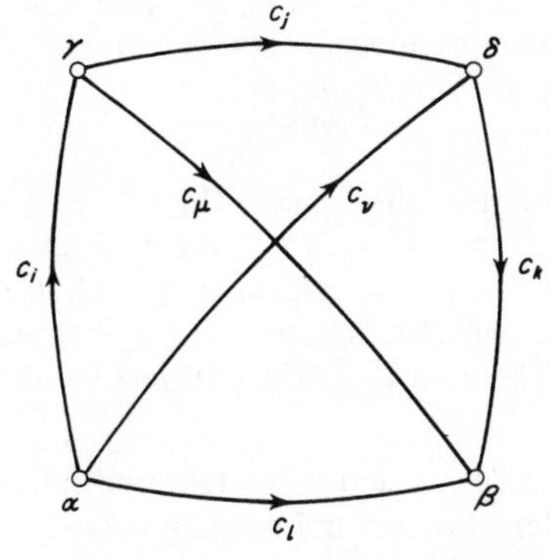

right side of the equation enumerates the same quadrangles, but by counting those with the diagonal (γ, β) coloured c_μ and summing.

As with any algebra of matrices the multiplication constants a_{ijk} can be computed if we know enough about the traces of the matrices in V, that is about the values of the character η.

Lemma 7 (i) $$\operatorname{tr}(B_i) = \begin{cases} n & \text{if } i = 0 \\ 0 & \text{if } i \neq 0 \end{cases}$$

(ii) $$\operatorname{tr}(B_i B_j) = \begin{cases} nn_i & \text{if } j = i^* \\ 0 & \text{if } j \neq i^* \end{cases}$$

(iii) $$\operatorname{tr}(B_i B_j B_{k^*}) = nn_k\, a_{ijk}.$$

Proof. The first assertion is trivial since $B_0 = I$ and since all the diagonal entries of B_i are 0 if $i \neq 0$. Then (ii) and (iii) follow directly from the description of the multiplication constants in the Corollary of Lemma 5. In fact (ii) and (iii) may also be seen combinatorially. For example, $B_i B_j B_{k^*}$ has as its (α, α)-coordinate the number of paths of length 3 from α back to α with steps coloured c_i, c_j, c_{k^*} in that order; since this number is $n_k\, a_{ijk}$ the trace of $B_i B_j B_{k^*}$ is $nn_k a_{ijk}$.

This last lemma is particularly useful when the multiplicities e_λ of the irreducible constituents of our representation of G by permutation matrices are all 1: then, as observed on page 102, a matrix U that completely reduces G will transform all the matrices in V to diagonal form and so the traces of products $B_i B_j$ or $B_i B_j B_{k^*}$ can be expressed simply in terms of the eigenvalues of their factors.

The algebra V does not easily distinguish primitive from general transitive groups. There is one small fact about eigenvalues however which does make this distinction:

Lemma 8. *Suppose that G is primitive on Ω, and that A is one of the adjacency matrices, whose corresponding subdegree is m where $m > 1$. Then*
(i) *m is an eigenvalue of A of multiplicity 1;*
(ii) *if θ is any other eigenvalue then $|\theta| < m$.*

The point here is the strict inequality in (ii). The matrix A may be a sum of some of the basic adjacency matrices, that is, the adjacency matrix of a monochrome subgraph of Ω obtained by amalgamating some of the original colours. The lemma is a special case of what is known as the Perron-Frobenius Theorem (see, for example, [14], Theorem 2, p. 65), but here is a direct proof of it.

Proof of Lemma 8. We consider A acting by right multiplication on the space K^n of 1 by n row vectors indexed by Ω (and we shall take K to be the complex field for simplicity). The vector $\mathbf{t} := (1, 1, \ldots, 1)$ is clearly an eigenvector with eigenvalue m since the column sums of A are all equal to m. Let $U := \{\mathbf{u} \mid \Sigma\, u_\alpha = 0\}$. Then $K^n = K\mathbf{t} \oplus U$, and since the row sums of A are all equal (to m) a simple calculation shows that U is invariant under A. Consequently it will be enough to show that if θ is an eigenvalue of A on U then $|\theta| < m$.

Now let $\mathbf{x} := (x_\alpha)$, and put

$$\mathbf{y} := (y_\alpha) := \mathbf{x}\,A.$$

If the (α, β)-coordinate of A is $m_{\alpha\beta}$ then

$$y_\beta = \sum_\alpha x_\alpha m_{\alpha\beta}$$

and

$$\|\mathbf{y}\|^2 = \sum |y_\beta|^2 = \sum_\beta \Big|\sum_\alpha x_\alpha m_{\alpha\beta}\Big|^2.$$

Now we use the fact that $m_{\alpha\beta}^2 = m_{\alpha\beta}$ (A is a (0, 1)-matrix) together with Cauchy's inequality:

$$\Big|\sum_\alpha x_\alpha\, m_{\alpha\beta}\Big|^2 = \Big|\sum_\alpha x_\alpha m_{\alpha\beta}^2\Big|^2$$

$$\leqslant \Big(\sum_\alpha |x_\alpha m_{\alpha\beta}|^2\Big)\Big(\sum_\alpha |m_{\alpha\beta}|^2\Big)$$

$$= m \sum_\alpha |x_\alpha|^2 m_{\alpha\beta}.$$

Hence

$$\|\mathbf{y}\|^2 \leqslant m \sum_\beta \sum_\alpha |x_\alpha|^2\, m_{\alpha\beta}$$

$$= m^2 \sum_\alpha |x_\alpha|^2$$

$$= m^2\, \|\mathbf{x}\|^2$$

and so $\|\mathbf{y}\| \leqslant m\,\|\mathbf{x}\|$. Returning to the applications of Cauchy's Inequality, equality holds if and only if, for every β the vectors $(x_\alpha m_{\alpha\beta})_{\alpha\in\Omega}$ and $(m_{\alpha\beta})_{\alpha\in\Omega}$ are linearly dependent. This means that for every β the coordinates x_α for which $m_{\alpha\beta} = 1$ are all equal; that is, $x_\alpha = x_\gamma$ if there exists β such that $(\alpha, \beta) \in \Delta$ and $(\gamma, \beta) \in \Delta$, where (Ω, Δ) is the graph corresponding to A; in other words, we require that $x_\alpha = x_\gamma$ whenever $(\alpha, \gamma) \in \Delta \circ \Delta^*$. Now $\Delta \circ \Delta^*$ is not empty since $m > 1$, and therefore it is a union of monochrome subgraphs of Ω. These are all connected since G is primitive (Lemma 3), and consequently $(\Omega, \Delta \circ \Delta^*)$ is connected. It follows that $x_\alpha = x_\gamma$ for all $\alpha, \gamma \in \Omega$, and so $\mathbf{x}$ is a multiple of $\mathbf{t}$. Thus, if $\mathbf{x} \notin K\mathbf{t}$ then $\|\mathbf{x}A\| < m\|\mathbf{x}\|$, and, as observed above, this proves the lemma.

The condition $m > 1$ in the lemma is no great restriction for, by Theorem 5, if G is a primitive group and one of its non-trivial subdegrees is 1, then G is a cyclic group of prime order acting regularly; in fact, for such a group the adjacency matrices are permutation matrices and their eigenvalues are roots of unity.

There is an easy converse of the lemma. For, if G is imprimitive then at least one of the monochrome subgraphs of Ω is disconnected, say with k components each of size l; if we take the elements of Ω in a suitable order then the adjacency matrix A of this graph will have zeros everywhere except in k square l by l blocks on its main diagonal, these blocks being the adjacency matrices of the components; then each block has m entries 1 in each row and column, and it is easy to see that A has k linearly independent eigenvectors with eigenvalue m.

To finish this lecture here are some notes relating my treatment of the centraliser ring to others in the literature.

Note 1. The description I have given of the centraliser ring is part of an account that I wrote in 1969 as a section of a paper on primitive groups of degree $3p$. This should appear ultimately in a modified form, incorporated as part of [29].

Note 2. The "intersection numbers" $\mu_{ij}^{(\alpha)}$ introduced by D. G. Higman (in [16], p. 29) are the same as our multiplication constants, as a comparison of Higman's definition with the discussion on pages 102, 103 above shows (see also [18], pp. 21, 22): our a_{ijk} is his $\mu_{ik}^{(j^*)}$. Higman introduces "intersection matrices" $M_\alpha := (\mu_{ij}^{(\alpha)})$ and observes that they span an algebra isomorphic to V. In fact, the definition of the numbers a_{ijk} shows that the intersection matrix is the matrix representing B_α in the regular representation of V and with respect to the standard basis of V.

Note 3. There is a natural one-one correspondence between suborbits of G in Ω and double cosets $G_\alpha x G_\alpha$ of the stabiliser G_α in G. For, $\{\alpha g \mid g \in G_\alpha x G_\alpha\}$ is an orbit of G_α in Ω, and different double cosets give rise to different suborbits. The subdegree corresponds to the number of cosets in the double coset; the pairing of suborbits corresponds to the pairing of a double coset $G_\alpha x G_\alpha$ with its inverse $G_\alpha x^{-1} G_\alpha$.

Now we can identify the centraliser ring V with the "double coset algebra", which is a subring of the group algebra KG: if $S_0, S_1, \ldots, S_{r-1}$ are the double cosets of G_α in G then the elements

$$\frac{1}{|G_\alpha|}\underline{S}_0, \quad \frac{1}{|G_\alpha|}\underline{S}_1, \ \ldots, \quad \frac{1}{|G_\alpha|}\underline{S}_{r-1}$$

(where, if $S \subseteq G$, then $\underline{S}$ denotes $\sum_{s \in S} s$ as element of KG) span a subalgebra of KG, and it is not hard to identify this algebra with V. It was in this form that the centraliser ring was first introduced by I. Schur [32] and studied by him and by J. S. Frame [9–12].

Note 4. If H is a regular subgroup of G (that is, if $G_\alpha H = G$ and $G_\alpha \cap H = 1$), and if $T_i := H \cap S_i$ where $S_0, S_1, \ldots, S_{r-1}$ are the double cosets of G_α in G (see Note 3), then the elements $\underline{T}_0, \underline{T}_1, \ldots, \underline{T}_{r-1}$ of the group algebra KH span an algebra, and again it can easily be identified with V. This is a so-called *Schur ring* or *S-ring* on H. Schur rings and their application to the study of permutation groups have received considerable successful attention (see, for example, [32]; [48], Chapter IV; [42] and references quoted there).

Note 5. There is another closely related theory, which arises in a very different context. If $G := SL(2, \mathbf{Q})$, the group of unimodular 2 by 2 matrices with rational coefficients, and $H := SL(2, \mathbf{Z})$, the subgroup of G consisting of matrices whose coefficients are integral, then, although H has infinite index in G, the conjugates of H intersect H in subgroups of finite index. Consequently the suborbits of G acting on the set of cosets of H are finite; equivalently, each double coset of H in G contains only finitely many cosets. It follows that we can still define the basic adjacency matrices as row-finite and column-finite matrices, which may be multiplied in the ordinary way. Thus our definition of the centraliser ring V can be generalised to this case. Here V is known as the *Hecke ring* (see, for example, [37], pp. 51–55), and this term is now used to describe double coset algebras by some authors, especially when the group G is an algebraic group of some kind (see, for example, [7]).

5. Applications of centraliser ring theory

As a general rule there does not seem to be much connection between the numerical data describing G as a permutation group on Ω and that describing G as a linear group, even though the permutation group determines the linear group so simply. There are however some weak relationships. Here is the first.

Theorem 6. *The highest common factor f of the degrees $f_1, f_2, \ldots, f_{s-1}$ of the non-principal irreducible constituents of G as linear group divides all the non-trivial subdegrees $n_1, \ldots, n_{r-1}$.*

Proof. We use the fact (Lemma 6) that if $i \geqslant 1$ then

$$n_i + \sum_{\lambda=1}^{s-1} f_\lambda \eta_\lambda(B_i) = \eta(B_i) = 0,$$

that is,
$$n_i/f = -\sum_{\lambda=1}^{s-1} (f_\lambda/f)\, \eta_\lambda(B_i).$$

Now f_λ/f is integral by definition of f; and $\eta_\lambda(B_i)$ is a sum of certain of the eigenvalues of B_i, which, since B_i is a matrix with integer entries, are algebraic integers; thus n_i/f is an algebraic integer. Since a rational number which is an algebraic integer must be an ordinary integer we see that f divides n_i.

Corollaries: (1) *If the non-principal character degrees f_i are all equal, say $f_1 = f_2 = \ldots = f_{s-1} = f$, then π is multiplicity-free and $n_1 = n_2 = \ldots = n_{r-1} = f$.*
(2) *If $r \leqslant 5$ then π is multiplicity-free.*

Here I have referred to the character π as being multiplicity-free, meaning of course the same as on page 102, that the multiplicities $e_1, \ldots, e_{s-1}$ are all 1 (and consequently $s = r$). The proof is as follows: we have that f divides n_i, so $f \leqslant n_i$; therefore

$$n = 1 + \sum_{\lambda=1}^{s-1} e_\lambda f_\lambda = 1 + \left(\sum_{1}^{s-1} e_\lambda\right) f \leqslant 1 + \left(\sum_{1}^{s-1} e_\lambda^2\right) f$$

$$= 1 + (r-1)f \leqslant 1 + \sum_{1}^{r-1} n_i = n.$$

Since the two inequalities must obviously be equality we have that $e_\lambda = 1$ for all λ and $n_i = f$ for all i. The second corollary follows immediately because if $r \leqslant 5$ then the only solutions of the equations $\sum_{0}^{s-1} e_\lambda^2 = r$, $e_0 = 1$, are (i) $s = r$ and $e_\lambda = 1$ for all λ, and (ii) $s = 2$, $e_0 = 1$, $e_1 = 2$. This latter possibility is excluded by Corollary 1.

A different proof of Corollary 2 is given by Burnside (in [2], Section 250, p. 338), and a significantly different proof of Corollary 1 is given by Wielandt (in [48], p. 92). It is a consequence of Clifford's Theorem in representation theory (see [8], Section 49, pp. 342–345) that if G arises as a normal subgroup of a doubly transitive group on Ω then the character degrees f_i are all equal. It follows that in this case π is multiplicity-free, an interesting and useful fact which was pointed out to me by Dr. J. P. J. Macdermott. The reverse of Theorem 6 and the converse of Corollary 1, that is, the statements that the highest common factor of $n_1, \ldots, n_{r-1}$ divides all the f_i, and that if $n_1 = \ldots = n_{r-1}$ then $f_1 = \ldots = f_{s-1}$, are false in general:

Examples. (i) If G is A_5 or S_5 acting on pairs from $\{1, \ldots, 5\}$ then G has rank 3, subdegrees 1, 3, 6 and character degrees 1, 4, 5.
(ii) If G is A_7 or S_7 acting on pairs from $\{1, \ldots, 7\}$, then G has rank 3, subdegrees 1, 10, 10 and character degrees 1, 6, 14.

Our next theorem is another arithmetic relationship between the subdegrees n_i and the character degrees f_i:

Theorem 7 (J. S. Frame, 1941). *If the permutation character π is multiplicity-free then $f_1 f_2 \ldots f_{r-1}$ divides $n^{r-2} n_1 n_2 \ldots n_{r-1}$. Moreover,*

$$n^{r-2} \frac{n_1 n_2 \ldots n_{r-1}}{f_1 f_2 \ldots f_{r-1}} = (-1)^t \, \Delta^2$$

where Δ is an integer in K, and t is the number of pairs of non-self-paired suborbits of G.

Proof. We first use Lemmas 6 and 7. Put

$$\theta_{\lambda i} := \eta_\lambda(B_i)$$

so that, since $e_\lambda = 1$, $\theta_{\lambda i}$ is an eigenvalue of B_i with multiplicity f_λ. The linear and quadratic trace relations of Lemma 7 can be written

$$n_i + f_1 \theta_{1i} + f_2 \theta_{2i} + \ldots + f_{r-1} \theta_{r-1\, i} = \begin{cases} n & \text{if } i = 0 \\ 0 & \text{if } i > 0 \end{cases}$$

$$n_i n_j + f_1 \theta_{1i} \theta_{1j} + f_2 \theta_{2i} \theta_{2j} + \ldots + f_{r-1} \theta_{r-1\, i} \theta_{r-1\, j} = \begin{cases} n n_i & \text{if } j = i^* \\ 0 & \text{if } j \neq i^*. \end{cases}$$

If T is the r by r matrix $(\theta_{\lambda i})$ then these equations simply say that

$$T^{\mathrm{Tr}} F \, T = n \, N,$$

where F is the diagonal matrix

$$\begin{pmatrix} 1 & & & & \\ & f_1 & & & \\ & & f_2 & & \\ & & & \ddots & \\ & & & & f_{r-1} \end{pmatrix}$$

and N is the diagonal matrix

$$\begin{pmatrix} 1 & & & & \\ & n_1 & & & \\ & & n_2 & & \\ & & & \ddots & \\ & & & & n_{r-1} \end{pmatrix}$$

multiplied by the matrix of the involution *. Taking determinants:

$$(-1)^t n^r \frac{n_1 n_2 \ldots n_{r-1}}{f_1 f_2 \ldots f_{r-1}} = (\det T)^2.$$

Now if we add f_λ times row λ to the first row of T $(1 \leqslant \lambda \leqslant r-1)$ we obtain a new matrix T_1 with the same determinant. But, from the linear trace relations it follows that

$$T_1 = \begin{pmatrix} n & 0 & \cdot\ \cdot & 0 \\ 1 & \theta_{11} & \cdot\ \cdot & \theta_{1r-1} \\ \cdot & \cdot & \cdot\ \cdot & \cdot \\ 1 & \theta_{r-11} & \cdot\ \cdot & \theta_{r-1r-1} \end{pmatrix}$$

and so $\det T = \det T_1 = n\Delta$, where $\Delta := \det(T_0)$ and T_0 is the $(r-1)$ by $(r-1)$ matrix $(\theta_{\lambda i})$. Since $\theta_{\lambda i}$ is an algebraic integer in K (because it is an eigenvalue of the integer matrix B_i) it follows that Δ is an algebraic integer, and since

$$(-1)^t \Delta^2 = n^{r-2} \frac{n_1 n_2 \ldots n_{r-1}}{f_1 f_2 \ldots f_{r-1}},$$

the result follows.

Notice that if the degrees $f_1, \ldots, f_{r-1}$ are all different then the rational field is a splitting field for G as a linear group, and furthermore all the suborbits are self-paired (see Theorem 8 below), and so Frame's quotient is a square integer. Frame has also proved [10] a generalisation of Theorem 7 applicable in case the multiplicities $e_1, \ldots, e_{s-1}$ are not all 1. The theorem can be used surprisingly often to exclude various possibilities for the n_i and the f_i in a hypothetical group G (cf., Ito[19]).

As a general rule information comes most easily when the representation of G as a linear group is multiplicity-free. We have already used this hypothesis in Theorem 7, it will be satisfied also in the last two theorems. Part of the point of the next theorem is that it gives a useful sufficient condition for multiplicity-freeness. From the description on pages 100, 101 of the simultaneous reduction of G and V it should be clear that V is isomorphic to the direct sum of e_i by e_i matrix algebras over the field K, and hence that G (or π) is multiplicity-free if and only if V is isomorphic to a direct sum of fields (isomorphic to K). *Thus π is multiplicity-free if and only if V is commutative.*

Theorem 8. *The suborbits of G are all self-paired if and only if π is multiplicity-free and every irreducible constituent χ_λ of π is real-valued.*

Proof. Suppose first that every suborbit is self-paired. Then all the basic adjacency matrices B_i are symmetric, so V consists of symmetric matrices; if $A_1, A_2 \in V$ then

$$A_1A_2 = A_1^{\mathrm{Tr}}A_2^{\mathrm{Tr}} = (A_2A_1)^{\mathrm{Tr}} = A_2A_1;$$

thus V is commutative and π must be multiplicity-free. Furthermore, since V consists of symmetric matrices it has the real number field $\mathbf{R}$ as a splitting field (that is, we can choose a real orthogonal matrix U so that $U^{-1}AU$ is diagonal for all A in V, and use U as on page 101), and then, since G as linear group spans the set of matrices centralising V, also G has $\mathbf{R}$ as a splitting field. This means that the absolutely irreducible constituents χ_λ of π are characters of representations realisable over $\mathbf{R}$, and so certainly each constituent χ_λ is real-valued.

For the converse suppose that $\pi = \chi_0 + \chi_1 + \ldots + \chi_{r-1}$, where the absolutely irreducible characters χ_λ are all different and are real-valued. The only difficulty in retracing the steps in the first part of the proof is to show that χ_λ is the character of a representation which can be realised over $\mathbf{R}$. Let $\mathbf{C}^n$ be the vector space of 1 by n row vectors with complex coordinates indexed by Ω, on which the group G (as group of permutation matrices) operates by right multiplication. There is then a unique subspace W_λ which is invariant under G, irreducible and affording the character χ_λ. Now if $\overline{W}_\lambda := \{\overline{\mathbf{x}} | \mathbf{x} \in W_\lambda\}$ then, since G consists of matrices with real entries, it is easy to see that $\overline{W}_\lambda$ is a G-invariant subspace of $\mathbf{C}^n$ and that it affords the character $\overline{\chi}_\lambda$. Since $\overline{\chi}_\lambda = \chi_\lambda$ it follows from the uniqueness of W_λ that $\bar{W}_\lambda = W_\lambda$. Consequently the subset of W_λ consisting of vectors with real coordinates is G-invariant, and is a vector space over **R** affording the character χ_λ. It now follows that the matrix U of page 100 can be taken to be real and orthogonal. Then since $U^{\mathrm{Tr}}B_iU$ is diagonal it follows that

$$B_i = U(U^{\mathrm{Tr}}B_iU)\,U^{\mathrm{Tr}} = U(U^{\mathrm{Tr}}B_iU)^{\mathrm{Tr}}U^{\mathrm{Tr}} = B_i^{\mathrm{Tr}}$$

for all i, and so every suborbit is self-paired.

The proof of Theorem 8 can be shortened a little if one uses some more technical character theory. One can also give a general formula for the number of self-paired suborbits in terms of the types and multiplicities of irreducible constituents of π. This, and more besides, is to be found in Frame's paper [10], expressed there in terms of double cosets.

My last illustration of the use of centraliser ring methods concerns primitive groups of degree kp where p is prime and k is small in some sense. The case $k := 1$ is a celebrated one about which much is known and more is not (see, for example, [27]), but it is not a good example for my present purposes. The story really begins with the analysis by H. Wielandt of primitive groups of degree $2p$ (see [47], or [48], Section 31, pp. 93–103). Using an old theorem of Jordan he proves first that if G is not alternating or symmetric then a Sylow p-subgroup P of G must be cyclic of order p. Then he exploits this fact to study the decomposition of the permutation character π by arguments similar to, but necessarily more complicated than, the treatment by Burnside of the permutation character of a group of prime degree. His result is that either

(i) $\pi = 1 + \chi_1$ with $f_1 = 2p - 1$

or

(ii) $\pi = 1 + \chi_1 + \chi_2$ with $f_1 = p - 1$, $f_2 = p$.

Thus a primitive group of degree $2p$ has rank at most 3. If its rank is 2, that is, if it is doubly transitive, then quite different methods are appropriate to find out more about it. But if the rank is 3 then centraliser ring methods are very suitable, so we will take the story up at this point.

We say that G is of *Wielandt type* (or sometimes, following L. L. Scott, of *2p-type*) if it is primitive, of rank 3 and with character degrees $1, m-1, m$. Thus $n = 2m$, and of the subdegrees n_1, n_2, one is odd and the other is even, so they are different and both the non-trivial suborbits must be self-paired. Since moreover the character π is rational valued, and its constituents χ_0, χ_1, χ_2 all have different degrees they too are rational valued. It follows easily that the rational field is a splitting field for G as a group of matrices, and therefore also for the centraliser ring V. The following lemma is very little different from what is to be found in [47]. A detailed analysis of rank 3 groups is to be found in D. G. Higman's work (see for example [15], [17]).

Lemma 9. *If G is of Wielandt type then there is a positive integer a such that*
(i) $m = 2a^2 + 2a + 1$ *(and so* $n = (2a+1)^2 + 1$*);*
(ii) *the subdegrees are* 1, $a(2a+1)$, $(a+1)(2a+1)$*;*
(iii) *G is a group of automorphisms of a regular graph of valency* $a(2a+1)$ *which has* a^2-1 *triangles on every edge.*

Proof. We may suppose that n_1 is the smaller of the two non-trivial subdegrees, so that $n_1 < m$, and we will use λ, μ for the other eigenvalues, of multiplicities $m-1$, m respectively, of B_1. Since these are eigenvalues of an integer matrix they are algebraic integers, and since they are rational λ and μ are therefore ordinary integers. Now we use the linear and quadratic trace relations of Lemma 7 as diophantine equations:

$$n_1 + (m-1)\lambda + m\mu = 0$$

$$n_1^2 + (m-1)\lambda^2 + m\mu^2 = 2mn_1.$$

From the linear equation, $n_1 \equiv \lambda \pmod{m}$. By Lemma 8, $|\lambda| < n_1$, and since $n_1 < m$ the only possibility is that $\lambda = n_1 - m$. Thus

$$n_1 = \lambda + m$$

and from the linear equation

$$\mu = -\lambda - 1.$$

Substitution in the quadratic equation gives

$$(\lambda^2 + 2\lambda m + m^2) + (m-1)\lambda^2 + m(1 + 2\lambda + \lambda^2) = 2m(\lambda + m)$$

and this simplifies to

$$2\lambda^2 + 2\lambda + 1 = m.$$

Now λ is negative, so if we put $a := \mu$, then a is a positive integer and $\lambda = -a - 1$,

$$m = 2(a^2 + 2a + 1) - 2(a+1) + 1 = 2a^2 + 2a + 1$$

$$n_1 = 2a^2 + a = a(2a+1)$$

$$n_2 = 2m - 1 - n_1 = 2a^2 + 3a + 1 = (a+1)(2a+1).$$

The number of triangles on each edge in the graph (Ω, Δ_1) is the multiplication constant a_{111} and this can be calculated from the cubic trace relation of Lemma 7:

$$a_{111} = \frac{a^3(2a+1)^3 - (2a^2+2a)(a+1)^3 + (2a^2+2a+1)a^3}{2(2a^2+2a+1)\,a(2a+1)}$$

which ultimately gives that $a_{111} = a^2 - 1$.

Of course all the other parameters of the coloured graph on Ω can now be calculated directly. Since (Ω, Δ_2) is the complementary graph of (Ω, Δ_1) we only need to describe the latter, which we do with the diagram:

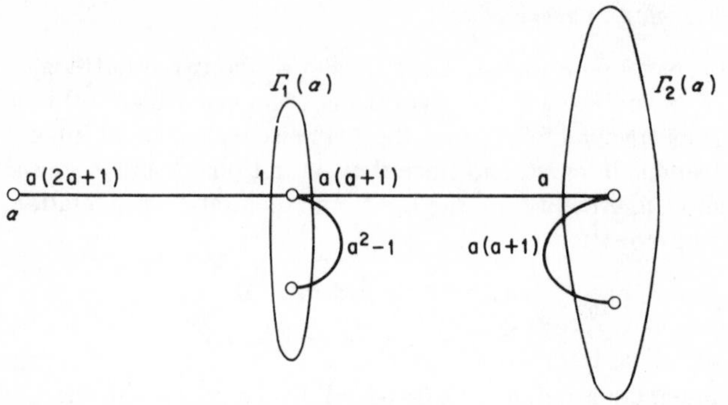

which indicates that of the $a(2a + 1)$ edges from each point of $\Gamma_1(\alpha)$ one returns to α, $a^2 - 1$ terminate in other points of $\Gamma_1(\alpha)$ and the remaining $a(a + 1)$ go to points of $\Gamma_2(\alpha)$. The fact that a^2 of the edges from a point of $\Gamma_2(\alpha)$ return to $\Gamma_1(\alpha)$ comes immediately from the knowledge that there are $a(2a + 1)\, a(a + 1)$ edges between $\Gamma_1(\alpha)$ and $\Gamma_2(\alpha)$, an instance of the relations mentioned on page 103.

This lemma illustrates the use of the strict inequality of Lemma 8 to exploit the primitivity of G: for every value of m there are imprimitive groups of degree $2m$ in which the character degrees are $1, m - 1, m$.

Exercise. Show that if G is imprimitive of degree $2m$ then G has rank 3 and character degrees $1, m - 1, m$ if and only if there is a system of imprimitivity with m blocks of size 2, such that the stabiliser of a block is transitive on its complement.

Returning now to groups of degree $2p$ we have

Theorem 9 (H. Wielandt, 1956). *If G is primitive of degree $2p$ where p is a prime number then either G is doubly transitive or it is a group of rank 3. This latter possibility can only occur if p is of the form $2a^2 + 2a + 1$, and then the subdegrees of G are 1, $a(2a + 1)$, $(a + 1)(2a + 1)$.*

The only known groups of Wielandt type are A_5 and S_5 of degree 10 acting on the set of pairs from $\{1, 2, 3, 4, 5\}$ (see p. 87 for a picture of the corresponding graph). L. L. Scott [33, 34] has shown that there are no more when $2a + 1$ is prime, and, partly in collaboration with G. Keller, he has shown that these are the only examples for $a \leqslant 12$, that is, for $n \leqslant 626$. Very recently he has also succeeded in proving that certain groups arising in a geometric and number-theoretic problem of Michael Fried are groups of Wielandt type (see [13] and [35]). The proof uses his theory of "orbital characters" (see [33]), which is a little beyond the scope of these lectures. It gives hope that the following problem should have an affirmative solution:

Problem 2. *Suppose that G is primitive of degree $2m$ and rank 3. Is it true that, if there is an element u of G which acts on Ω as a product of two disjoint m-cycles, then G is of Wielandt type?*

If this is true then one should gain some more insight into a problem of Mathieu concerning groups of prime degree (see [27], p. 525, or [28]).

An analysis of primitive groups of degree $3p$ has been completed and is to be published shortly as joint work of L. L. Scott, Olaf Tamaschke and me. The strategy is much the same as Wielandt's for groups of degree $2p$, but the calculation is significantly harder. One quickly reduces the problem to studying a primitive group G of degree $3p$ whose Sylow p-subgroups are cyclic of order p (indeed, this reduction can be done for groups of degree kp with $k < p$). Then one uses theorems of R. Brauer and W. Feit about the ordinary and modular irreducible representations of a group with a cyclic Sylow p-subgroup to analyse the permutation character of G. The result is a list of four or five possibilities for its decomposition, and these are then treated using the centraliser ring. The calculation amounts to using the linear and quadratic trace relations of Lemma 7 as diophantine equations in the eigenvalues of the basic adjacency matrices as in the proof of Lemma 9, but since the rank may be 4 it is very much harder to complete and uses very much more of one's detailed knowledge of the centraliser ring. The result is

Theorem 10 (ΠMN, L. L. Scott and Olaf Tamaschke). *A simply primitive group of degree $3p$ where p is prime can exist only if either*

(i) $p = 192a^2 + 60a + 5$ *or*
(ii) $p = 3a^2 + 3a + 1$

for some integer a. Furthermore the rank is at most 4, and there are only a few possibilities for the character decomposition and the corresponding centraliser ring.

In each case we have worked out all the eigenvalues of the basic adjacency matrices and so, by Lemma 7, we can tabulate complete numerical information about the coloured graph of which G is an automorphism group.

David Cooper has carried out a similar calculation for groups of degree $4p$, which is the content of his Oxford D.Phil. thesis (1976). There is only one case which he has not yet completed: that where the group has rank 5 with all its suborbits self-paired, and where the degrees of the irreducible constituents of π are $1, p-1, p, p, p$. For larger values of k the groups of degree kp seem to be impossible to handle by this method, and if progress is to be made the centraliser ring calculations will need to be supplemented with some essentially new ideas.

References

1. Elizabeth R. Bennett. Primitive groups with a determination of the primitive groups of degree 20. *Amer. J. Math.* **34**(1912), 1–20.
2. W. Burnside. "Theory of Groups of Finite Order" 2nd Edition, Cambridge University Press, 1911 (reprinted by Dover Publ. Inc., New York, 1955).
3. P. J. Cameron. Proofs of some theorems of W. A. Manning. *Bull. London Math. Soc.* **1**(1969), 349–352.
4. P. J. Cameron. Permutation groups with multiply transitive suborbits, I, II. *Proc. London Math. Soc.* **25**(3) (1972), 427–440 and *Bull. London Math. Soc.* **6**(1974), 136–140.
5. P. J. Cameron. Another characterization of the small Janko group. *J. Math. Soc. Japan* **25**(1973), 591–595.
6. P. J. Cameron. Suborbits in primitive groups. *In* "Combinatorics", Part 3, pp. 98–129 (Edited by M. Hall Jr. and J. H. van Lint), Mathematical Centre Tracts **57**, Amsterdam, 1974.
7. C. W. Curtis, N. Iwahori, and R. Kilmoyer. Hecke algebras and characters of parabolic type of finite groups with (B, N)-pairs. *I.H.E.S. Publ. Math.* **40** (1971), 81–116.
8. C. W. Curtis and I. Reiner. "Representation Theory of Finite Groups and Associative Algebras". Interscience (John Wiley and Sons), New York, 1962.
9. J. Sutherland Frame. The degrees of the irreducible components of simply transitive permutation groups. *Duke Math. J.* **3**(1937), 8–17.
10. J. Sutherland Frame. The double cosets of a finite group. *Bull. Amer. Math. Soc.* **47**(1941), 458–467.
11. J. Sutherland Frame. Double coset matrices and group characters. *Bull. Amer. Math. Soc.* **49**(1943), 81–92.
12. J. Sutherland Frame. Group decomposition by double coset matrices. *Bull. Amer. Math. Soc.* **54**(1948), 740–755.
13. M. Fried and D. J. Lewis. Solution spaces for diophantine problems. To be published in *Bull. Amer. Math. Soc.*
14. F. R. Gantmacher. "Applications of the Theory of Matrices". Interscience (John Wiley and Sons), New York, 1959.

15. D. G. Higman. Finite permutation groups of rank 3. *Math. Zeitschrift* **86** (1964), 145–156.
16. D. G. Higman. Intersection matrices for finite permutation groups. *J. Algebra* **6**(1967), 22–42.
17. D. G. Higman. A survey of some questions and results about rank 3 permutation groups. *In* "Proc. Internat. Congress of Mathematicians, Nice 1970", Vol. 1, pp. 361–365, Gauthier-Villars, Paris, 1971.
18. D. G. Higman, "Combinatorial considerations about permutation groups". Lecture notes, Mathematical Institute, Oxford, 1972.
19. Noboru Ito. Transitive permutation groups of degree $p = 2q + 1$, p and q being prime numbers, II, III. *Trans. Amer. Math. Soc.* **113**(1964), 454–487 and **116**(1965), 151–166.
20. Camille Jordan. "Traité des Substitutions et des Équations Algébriques". Gauthier-Villars, Paris, 1870.
21. Wolfgang Knapp. On the point stabilizer in a primitive permutation group. *Math. Zeitschrift* **133**(1973), 137–168.
22. Wolfgang Knapp. Primitive Permutationsgruppen mit einem zweifach primitiven Subkonstituenten. *J. Algebra* **38**(1976), 146–162.
23. Donald Livingstone. On a permutation representation of the Janko group. *J. Algebra* **6**(1967), 43–55.
24. W. A. Manning. "Primitive Groups" (Part 1). Stanford University Publications, Vol. 1, 1921.
25. W. A. Manning. Simply transitive primitive groups. *Trans. Amer. Math. Soc.* **29**(1927), 815–825.
26. W. A. Manning. A theorem concerning simply transitive primitive groups. *Bull. Amer. Math. Soc.* **35**(1929), 330–332.
27. Peter M. Neumann. Transitive permutation groups of prime degree. *In* "Proc. Conf. Theory of Groups, Canberra 1973". Springer Lecture Notes, **372**(1974), 520–535.
28. Peter M. Neumann. Transitive permutation groups of prime degree, IV: a problem of Mathieu and a theorem of Ito. *Proc. London Math. Soc.* (3)**32** (1976), 52–62.
29. Peter M. Neumann, Leonard L. Scott, and Olaf Tamaschke. Primitive permutation groups of degree $3p$. To be published.
30. William L. Quirin. Extension of some results of Manning and Wielandt on primitive permutation groups. *Math. Zeitschrift* **123**(1971), 223–230.
31. Henry Lewis Rietz. On primitive groups of odd order. *Amer. J. Math.* **26**(1904), 1–30.
32. I. Schur. Zur Theorie der einfach transitiven Permutationsgruppen. *Sitzungsber. Preuss. Akad. Wiss. Berlin* (1933), 598–623 = Gesammelte Abhandlungen, Springer Verlag, 1973, III, 266–291.
33. Leonard L. Scott. "Uniprimitive groups of degree kp". Ph.D. dissertation, Yale University, 1968.
34. Leonard L. Scott. On permutation groups of degree $2p$. *Math. Zeitschrift,* **126**(1972), 227–229.

35. Leonard L. Scott. On a problem of M. Fried. *In* "Proc. Conf. Finite Groups, Utah 1975" (Edited by W. R. Scott and Fletcher Gross), Academic Press, New York, 1976.
36. J.-A. de Séguier. "Éléments de la Théorie des Groupes de Substitutions". Gauthier-Villars, Paris, 1912.
37. Goro Shimura. "Introduction to the Arithmetic Theory of Automorphic Functions". Publ. Math. Soc. Japan, Vol. 11, Iwanami Shoten and Princeton University Press, 1971.
38. Charles C. Sims. Graphs and finite permutation groups. *Math. Zeitschrift* **95**(1967), 76–86.
39. C. C. Sims. Graphs and finite permutation groups, II. *Math. Zeitschrift* **103**(1968), 276–281.
40. C. C. Sims. Computational methods in the study of permutation groups. *In* "Computational Problems in Abstract Algebra" (Edited by J. Leech), Pergamon Press, Oxford, 1970, pp. 169–183.
41. C. C. Sims. Primitive groups, graphs and block designs. *Annals New York Acad. Sciences* **175**(1970), Art. 1, 351–353.
42. Olaf Tamaschke. *S*-rings and the irreducible representations of finite groups. *J. Algebra* **1**(1964), 215–232.
43. John G. Thompson. Bounds for orders of maximal subgroups. *J. Algebra* **14**(1970), 135–138.
44. W. T. Tutte. A family of cubical graphs. *Proc. Camb. Phil. Soc.* **43** (1947), 459–474.
45. W. T. Tutte. On the symmetry of cubic graphs. *Can. J. Math* **11**(1959), 621–624.
46. Marie J. Weiss. On simply transitive groups. *Bull. Amer. Math. Soc.* **40**(1935), 401–405.
47. Helmut Wielandt. Primitive Permutationsgruppen vom Grad 2*p*. *Math. Zeitschrift* **63**(1956), 478–485.
48. Helmut Wielandt. "Finite Permutation Groups". Academic Press, New York and London, 1964.
49. Helmut Wielandt. "Permutation groups through invariant relations and invariant functions". Lecture notes, Department of Mathematics, The Ohio State University, 1969.
50. Helmut Wielandt. "Subnormal subgroups and permutation groups". Lecture notes, Department of Mathematics, The Ohio State University, 1971.
51. Warren J. Wong. Determination of a class of primitive permutation groups. *Math. Zeitschrift* **99**(1967), 235–246.